GENERAL VIEW OF THE
AGRICULTURE OF
OXFORDSHIRE

GENERAL VIEW of the AGRICULTURE of OXFORDSHIRE

A Reprint of the Work Drawn up for the Consideration of the Board of Agriculture and Internal Improvement

by

ARTHUR YOUNG

Secretary of the Board

AUGUSTUS M. KELLEY, PUBLISHERS
NEW YORK 1969

Published in the United States of America by
Augustus M. Kelley, Publishers, New York

SBN 678 05543 2
Library of Congress No 74-91998

This book was first published in 1813

This edition published in 1969

Printed in Great Britain by
Clarke Doble & Brendon Limited Plymouth

GENERAL VIEW

OF THE

AGRICULTURE

OF

OXFORDSHIRE.

DRAWN UP FOR THE CONSIDERATION OF

THE BOARD OF AGRICULTURE

AND INTERNAL IMPROVEMENT.

BY

THE SECRETARY OF THE BOARD.

LONDON:

PRINTED FOR SHERWOOD, NEELY, AND JONES,

PATERNOSTER-ROW :

SOLD BY G. AND W. NICOL, PALL-MALL; PARKER, COOKE, AND

MUNDAY, OXFORD; RUSHER, BANBURY; AND

J. RUSHER, READING.

1813.

[*Price Twelve Shillings in Boards.*]

London: Printed by B. M'Millan,
Bow Street, Covent Garden.

ADVERTISEMENT.

THE desire that has been generally expressed, to have the AGRICULTURAL SURVEYS of the KINGDOM reprinted, with the additional Communications which have been received since the ORIGINAL REPORTS were circulated, has induced the BOARD OF AGRICULTURE to come to a resolution to reprint such as appear on the whole fit for publication.

It is proper at the same time to add, that the Board does not consider itself responsible for every statement contained in the Reports thus reprinted, and that it will thankfully acknowledge any additional information which may still be communicated.

N. B. *Letters to the Board, may be addressed to Sir* JOHN SINCLAIR, *Bart. M. P. the President, No. 32, Sackville-Street, Piccadilly, London.*

PREFACE.

THE anxiety of the PRESIDENT of the Board to have the Reports of the English Counties finished, induced him to propose my examining a county in the summer of 1807 ; but having already executed that task in five counties, and being sensible that such works are more advantageously performed by various persons than by a few, united with the resolution that Essex should be my last labour in this way, I begged him to excuse me : difficulties, however, having occurred in procuring a Surveyor for Oxfordshire, that county alone remained unassigned, and the PRESIDENT again applied to me to undertake it ; and finding him much set on completing the kingdom, I complied with his request, however contrary it was to my own judgment and inclination.

A hope had long been entertained, that it would have been executed by D . WILLIAMS, Professor of Botany in the University of Oxford : had the undertaking been so fortunate as to have been placed in such able hands, it would have been executed in a manner more to the satisfaction of the Board.

OXFORD.]   I have

I have too often addressed the readers of the Reports, to render an explanation of the principles by which I am governed in these works necessary; and have therefore only to request, that they will bear in mind the very material object of that GENERAL REPORT, in which is designed to be conveyed a concentrated view of the National Agriculture; the practices that deserve commendation; the deficiencies that demand improvement; the means of spreading whatever is good; and of remedying whatever is defective. That work can only be built on the minute details of the various Surveyors; whose facts, though too numerous for entertainment, will not be useless in the hands of those, whose business it will be to reduce the whole to order and consistency.

Upon the circumstance of repetitions, which at first sight may appear too numerous, one observation should be made: suppose that in the quantity of seed, time of sowing, or any similar point of management there is a very near resemblance, I have heard it remarked, that one line would contain the matter as well as a page: thus,

Seed wheat two bushels.

But it may be asked, is such a general average, that of all farmers indiscriminately? or only of good ones? or of those the Reporter conversed with, whether good or indifferent managers? If variations, do they tend to lessen or increase it?

How

How far does the practice of the best correspond with that of the rest? And then, who are what the Reporter calls *good?* Who knows that they really are so, if not named? Hence the satisfaction of finding the authority every where named. Without minutiæ of this sort, the reader, from the first page to the last of a book, has always the writer between him and the farmer, and not as a transcriber, but as a calculator of effects or averages. I know not how other readers may feel, but to myself, in examining the description of a district, I wish to see authority named for all that is possible; as it is not only a proof that the writer has had such conversations on the spot, but that there exist sufficient judges resident in the country, who know whether the farmers named were, or were not, proper men to apply to.

CONTENTS.

CHAP. XIII. LIVE STOCK.

CHAP. XIV. RURAL ECONOMY.

CHAP. XV. POLITICAL ECONOMY.

CHAP. XVI. OBSTACLES TO IMPROVEMENT, 341

AGRICULTURAL SURVEY

OF

OXFORDSHIRE.

CHAP. I.

GEOGRAPHICAL STATE AND CIRCUMSTANCES.

SECT. I.—SITUATION AND EXTENT.

" THE county of Oxford extends in length, from the north-west extremity to the south-east, 50 miles. Being of a very irregular figure, it is not above seven miles across near the middle, at Oxford; though in the more northern part of the county, it measures 38 miles in diameter. The southern end is also narrow, being not more than 12 miles across in any part south of Oxford.

" It contains fourteen hundreds, one city, twelve market-towns, 207 townships or parishes, and about 450,000 acres; of which the part north of Oxford contains 309,000 acres, and the part south of Oxford 141,000 acres, as appears by the Topographical Survey which I have taken."

In the Table of Poor's-rates drawn up under the in-

 spection

spection of the Right Hon. George Rose, the contents
are stated at 474,880 acres.

" Oxfordshire is bounded on the south, south-west,
and south-east, by the county of Berks; on the east
by the county of Bucks; on the north-east by the
county of Northampton; on the north-west by the
county of Warwick; and on the west by the counties
of Worcester and Gloucester."

SECT. II.—DIVISIONS.

Oxfordshire is divided into the hundreds of

Banbury,	Thame,
Bloxham,	Dorchester,
Ploughley,	Lewknor,
Chadlington,	Pirton,
Wotton,	Ewelm,
Bullington,	Langtree,
Bampton,	Binfield.

The whole is in the diocese of Oxford.

SECT. III.—CLIMATE.

" The climate of Oxfordshire may be accounted, in
general, cold, particularly the westward part of the
northern division, where the fences consist chiefly of
stone walls, and consequently afford little or no shelter.
It is cold, also, upon and near the Chiltern hills, especially on the poor white lands at the foot of the hills;

                                                    where

where it is always to be observed, that the frost will take effect sooner, and continue longer on that soil, than it does on the deeper lands farther situated from the hills. *The climate of the Chiltern country is moist, on account of the fogs, which are more frequent on the hills and woods than in the vale.*"

SECT. IV.—SOIL.

THIS county contains three distinctions of soil, that are so marked by Nature, as to allow of little doubt respecting them.

I. The red-land of the northern district; which, in fertility, much exceeds that of any other portion of equal extent.

II. The district of stonebrash.

III. The Chiltern hills. These are defined by their outlines, so as to render the annexed Map, I trust, satisfactory. The remainder of the county is so various, and the different soils so intermixed, and the contiguous tracts of each so small, that they may be classed together under the title of,

IV. Miscellaneous loams. The proportionate extent of these soils, taking the total of the county at 450,000 acres, may thus be stated in the estimation of Mr. Neele, Map Engraver to the Board:

	Acres.
Red-land	79,635
Stonebrash	164,023
Chiltern	64,778
Miscellaneous	166,400
Total	474,836

I. The

I. The Red-land District.

Very nearly the whole range of country 13 miles, from Banbury to Chipping Norton, is enclosed by Act of Parliament, and improved in product very greatly. Much in grass, but much, also, arable. It is all red-land on gritstone till within three miles of Chipping Norton, where the yellow lime stonebrash begins.

The soil in the more northern part of the county is the rich red loam and sandy on a red gritstone rock, which they break for the turnpike-roads, of which it makes execrably bad ones, but the stonebrash good.

Wardington, Bourtons, Huscots, Cropedy, Horley, Hornton, Wroxton, Drayton, Hardwick, Hanwell, all red; Enwell, red sand; the Tadmertons, sandy; Milcomb, Bloxham, deep red loam; Claydon, red clay.

All the above are in the red-land district, enclosed chiefly by Act of Parliament. The soil uncommonly good, and letts in general, tithe-free, at 40s. per acre. There are some exceptions; but a finer district of soil is not often to be met with, whether in grass or arable.

Bloxham.—The field is amongst the finest lands in Oxfordshire, and qualitied by the Commissioners at 40s. an acre round.

There is at Atterbury, a very remarkable instance of the former ignorance of Commissioners of Enclosure: there are two soils—clay, and an admirable red sand; and when the parish was enclosed, the valuation was the grossest blunder that could be made; for the clay was valued at 12s. to 14s. per acre higher than the sand: whereas at present, every one knows the sand to be far more valuable than the clay. It was done many years ago. The sand would lett, through many parts of this kingdom, at 40s. per acre, and in some at 3l.

Sibberd

Sibberd Ferris, red sand; Burdrop, ditto; hills sand, bottom clay; Hook Norton, various, much deep red; Wiggington, red, some light sand; Little Compton, white stonebrash, and very good; Little Rolwright, red.

This red district, in respect of soil, may be considered as the glory of the county. It is deep, sound, friable, yet capable of tenacity; and adapted to every plant that can be trusted to it by the industry of the cultivators. It extends into Northamptonshire, Warwickshire, and will become an object to be traced by the Surveyors of the Board, I believe, from Devonshire to Nottinghamshire.

II. The Stonebrash District.

In tracing the outline of this district, we may begin on the borders of Gloucestershire, at Broughton Poggs, Broadwell, Alrescot, and Black Bourton; thence to Brise Norton, Witney, taking in Blenheim and Woodstock; leave out Bletchington; include Chesterton; cross Bicester; include Fringford, and leave out Newton Purcel; steer north to Shelswell, and take a part of Mixbury; and then in a south-west direction take in Cotsford, Fritwell, part of Somerton, and part of North Aston; leave out Dun's Tew, and part of Westcot Barton: then by Church Enston; leave ou Sandford; pass between Little Tew and Heathorp to Swaford; take in Great Rollright, and strike the bounds of Warwickshire at Little Rollright.

The predominant feature of this extensive tract is, a surface of greater or less depth, of a loose, dry, friable sand or loam, apparently formed of abraded stone, and abounding with many fragments of it. This is generally limestone; in every place that I viewed, it is so,

and

and the uniform intelligence was to the same purport. It forms an excellent turnip soil, and is productive in wheat. Modern rents rise to 20*s.* per acre, tithe-free; and much rises to 25*s.*

Bignal is stonebrash; and on Mr. Forster's farm I found, that in these limebrash lands there are many springs, which spew up upon the slopes, and greatly injure large tracts of them; and in digging Mr. Forster's capital drains four feet deep, there is found under the surface, brash of 12, 18, and 24 inches deep, a bed of blue clay apparently, which digs up very firm, and hardens in the air; but by trial with vinegar, I found it a marl. Under this is a vein of white marl, very rich in calcareous earth, as I judge, from its effervescence with acids; under this is a rock of rough white limestone. These blue and white substances, Mr. Forster has spread on the surface-stonebrash, and found them very beneficial.

In the vales, various loams are found, and some tracts of good meadow. Of course, in so large a tract of country, many variations are met with; from poor, loose, sandy, slopes, to deep and more heavy soils, which have been called (but I conceive erroneously) clays. Enclosure has flourished greatly throughout this district; but much yet remains open, a field for great improvement.

Burford, enclosed 13 years ago, is in general a stonebrash; but there was a large tract of heath-land, which is still of a more loose and hollow quality, and which demands a more attentive management. On this land, the layers are always pared and burnt; but not on the brash, because too stony for the operation.

Minster Level belongs to Mr. Coke, of Norfolk. It is a tract of turnip land, which Mr. Pratt thinks lett from

from 15*s.* to 20*s.* per acre. It is much intermixed with woods; and *therefore* perhaps it is, in his opinion, that they are subject to the mildew.

III *The Chiltern District.*

The basis of this tract of country is chalk ; in some places very white, and to the eye, pure ; in others imperfect, but in all consisting, in much the greater part, of calcareous earth. The chalk is covered to various depths with loam, generally sound and dry, and well adapted to beech-woods and sainfoin, which may be considered as the more peculiar products of the whole district. The clay which forms this loam abounds more or less, and but rarely takes any character that permits an arrangement with argillaceous soils. The most distinguishing mark of the surface-loam is, a very considerable quantity of flints ; mostly brown, rough, rusty, and honey-combed ; many to perforation, and many, also, with a sparry incrustation. Some of these I found, that effervesced with spirit of salt, as I expected that all would have done ; but was disappointed in various trials. Some of the best soils are the most covered with these flints ; so that strangers, unacquainted with soils, are apt to think that land must be miserably poor which is worth 20*s.* per acre.

On the banks of the Thames, and in some other places where the hills recede, the soil is an excellent sandy loam, free from these peculiarities, and yielding many turnips ; a crop which also do well on the flinty hills.

The bounds of this district extend along the Ickenfield hills, from Chinner to the vicinity of the Thames at Gatehampton.

IV. The

IV. The District of Miscellaneous Loams.

Having marked off, as in the Map, the three preceding districts, there remains a large portion of the county to which no particular characteristic can be assigned : it includes all sorts of soils, from loose sand to heavy clay.

Descending the Chiltern hills, from Stokenchurch to Tetsworth, the vale is open field, and the soil exceedingly good ; a brown strong loam on a moist bottom, yields great crops of wheat.

Milton field is one of the finest soils I have met with in the county : a dry, sound, friable loam on gravel— convertible land, as they call it in Oxfordshire ; but I know not why, unless that it is generally good for every thing : it would not be converted into grass, were it enclosed. It was covered with the finest show of turnips that I have seen in this failing year (1807), and among them a great breadth of Swedes, very fine and luxuriant.

Crossing from Thame to Postcomb, and thence by Stoke to Golder, I passed much good pasturage, and some very fine open-field arable.

At Golder, Mr. Cozins has three sorts of land ; first, a cold-bottomed land, upon which turnips cannot be had without destruction to the barley ; it is a whitish calcareous loam, which is apt easily to run together with rain, and then to cake and incrust. Secondly, gravel in clay, very stiff, tenacious, and difficult to plough, and when hard, mattocks will scarcely pick it. Thirdly, what he calls ragstone rubble, but improperly : it is chalk, and apparently good chalk, but with a mixture of clay, here called loam. It is found on the sides of hills, in strata of different thickness, which breaks

by

by the air into flakes, and other forms, but not any general fatting into powder. It becomes what is here called a clay soil ; in other counties, a *malmy* one.

At Stoke Talmage, much good grass, and the arable a good loam.

Golder.—Quitting the low land of Stoke, the hills of Golder are strong land ; much chalk loam, which breaks into flakes in the subsoil. Much open-field intermixed, from Golder to Bensington. The soil, part calcareous, and towards Bensington a loose turnip gravel. At Crowmarsh, a good gravelly loam in the vale. The hills are white calcareous loam, on a chalk loam rock. Excellent land for sainfoin, but bad for clover.

At Mungwell, the Bishop of Durham's farm is chiefly the same calcareous hilly land, upon a rough, coarse, chalkstone bottom. Better barley than wheat land. The parish of Mungwell, like some others on the banks of the Thames, consists of several belts of soil, rising from the river to the beech-woods which crown the hills. Upon the banks of the river, there is a tract of rich meadow. A gentle slope of loam on chalk, still arising, a third tract of the same soil, but with many flints. This is stronger land than the preceding ; and at last, the region of wood, intermixed with many fields of clay and flints, still stronger than the third belt. The breadth of these lines of soil varies according to the slope, and other circumstances of the country ; and the original distribution of the parishes seems to have been made, with a view to each having water at one end, and wood at the other.

At English, upon the farm of Mr. Dean, there are several strong specimens of the soil of the Chiltern hills —strong and difficult of tillage, yet dry enough for
turnips,

turnips, and abounding with red and white flints to a great degree. I saw some spots so covered with them, that I question whether earth was visible on a fiftieth part of the surface. Yet such is good corn land : and the finest crop of turnips I have seen this year (1807), was in a field remarkably full of them.

From English to the descent from the hills into the vale at Whitechurch, there is a great range of beech-woods, intermixed with chalk and flint fields : the same calcareous soil of loam and stone which has so often occurred in the Chiltern district. The breaches in the hills that front the vale of the Thames, every where discover a deep stratum of chalkstone, much of which appears to be pure, as far as can be judged from its whiteness. From Mapledurham to Caversham, and from Caversham to Henley, is a very fine tract of good sandy loam, under good management, and well sprinkled with good turnips. The neighbouring hills are various ; some of them seeming to be gravel, while others shew the chalk.

All the way to Stonor, the country gradually rises, and is in general the flinty soil of the Chiltern hills.

In the district of Culham, Clifton, Burcot, Dor-chester, and part of Warborough, there are several sorts of soil ; gravel, red sand, red sand upon gravel; these are the most general. Red clay, but not wet, because under-drained, and then dry enough for tur-nips; a friable, deep, good loam; a springy gravel, with a red clay under it : but of these soils, the red sand and the deep loam most deserve attention. It is a new instance to be added to the thousands that occur in every part of the world, of the fertility attending a red colour. From shallowness on gravel, and from a mixture of gravel in the sand, many fields, and parts

of

of fields are liable, in dry seasons, to burn; but no sand existing, black, yellow, or white, would, in equal circumstances, be equally valuable. On Culham-heath, the reddest sand (near Nuneham-lodge) is covered with thick fern—a sure proof every where of what is below it. Where the soil changes, the fern disappears, and the far inferior plant of furze takes it place. Much red sand at Marsh Baldon; Warborough, red sandy gravel; Drayton, red; Newington and Chalgrove, grey; Chisslehampton, strong; Toot Baldon, wet; Nuneham, mostly red sand; Sandford, strong land; Littlemoor, red sand; Horsepath, wet; Gasington, wet.

Burcot, Dorchester, Clifton—Gravelly loam, convertible.

Culham-field—Open, but laid together: sand, and sandy loam, convertible to any purpose.

Culham-heath—Allotted by Sir Cecil Bisshopp.

Chisslehampton—Much good grass under dairies: breeds mixed.

Stradhampton, Ascot, Brookhampton—Good arable enclosed, convertible: some meadow.

Newington—All grass.

Drayton—Open-field, gravelly, convertible: some Lammas grass lands.

Warborough—Flinty gravel.

Miltons—Very fine gravelly loam, open; subject to the mildew. Some stonebrash.

Great Hasely—A sort of stonebrash: all open-field arable, in the old system: some turnips, but by no means general.

Chalgrove—Clay; sad roads, and bad husbandry: all open.

Nuneham Courtney—Part clay, but the upper lands sand: some meadow. The beautiful park, decorated

with

with so much taste, contains above 1100 acres, part arable.

Sandford—Clay, enclosed : both grass and arable.

Garsington—Clay ; the uplands stonebrash : all open, except grass, of which not a little.

Horsepath—Stonebrash, open.

Cuddesdon—Sand, enclosed ; some good dairy land ; one or two complete dairies : butter and suckling.

Wheatley—Much sandstone, open.

Waterperry—All grass : dairies and suckling.

Waterstock—All grass : dairy. Lies on the Thame.

Aldbury, Tiddington—Enclosed : grass under dairies.

Holton—Enclosed : arable, but some rich pastures.

Stanton St. John—Enclosed arable.

Horton—Clay : pasture, and some arable.

Forest-hill—Enclosed : pasture : cows.

Shotover—Enclosed : arable and pasture. There is much yellow ochre here, and red sand.

Beckley—Open : pasture and arable.

Eldsfield—Clay, and some sand and stonebrash : all enclosed.

Church Cowley—Sand, arable.

Iffley—Sand, open.

Headington—Sand, and some stonebrash : enclosed arable.

Marston—All under cows.

Wood Eaton—Rich strong land : cow pastures and arable.

Water Eaton—All enclosed : rich cow-ground : some arable.

Kidlington—Cow-ground and arable.

There is a tract of reddish sand, three miles wide and four or five miles long, in the parishes of Beckley (where it lies on a rock), Eldsfield, Heddington, Shotover,

over, Stanton, and Forest-hill, upon which turnips succeed greatly. This tract of sand is found also in the upper part of Cowley-field, where it is deep and good; and in Little-moor, it is a famous deep sand. Marsh Baldon is a red sand. *Quere*—therefore, if the above tract does not unite with the fine red sands of Nuneham and Dorchester hundred?

At North-Weston, in the rich district of Thame, about one-third is under the plough. The grass lands may be divided into upland, and low flat meadows on the river. The land, a very fine sandy loam, or rather a loamy sand; but there are tracts of wet adhesive loam on clay, but very rich. The red, light, sandy loam is more predominant. Thame open-field is a reddish loam on a flinty gravel, and is an admirable soil. Morton-field, in the same district, a stiff loam: two crops and a fallow.

Kidlington—All open-field, and has a large cow-common.

Weston-on-the-Green—Enclosed: stonebrash.

Shipton—The same.

Water Eaton—Enclosed: cows and sheep.

Bampton, and its hamlets of Haddon, Weald, Cote, Aston, Lew, Chimney, and Shifford, contain about 6000 acres of arable, and 4000 of grass. Rich, deep, brown loam on gravel; part of it clay, on the hills; but the vale exceeding fine land, from 30*s.* to 40*s.* per acre. Much of it is thrown up into broad arched lands, and too high for the quality of the soil. Gravel seems to be under all the country, from Whelford and Lechlade to Bampton.

Kemscott—On gravel, and deep.

Grafton—Ditto.

Alrescot

Alrescot—On gravel, and stonebrash.

Clanville—Gravel and good loam.

Chimley—Good loam on gravel: excellent. Enclosed: no fallow.

Stanlake—Gravel.

Bright Hampton—On ditto.

Hardwick—Ditto.

Duckleton—Ditto, and clay.

Coggs and Ensham—Loam and clay.

Stanton Harcourt and North-moor—Gravel.

Newland—Stone and clay.

South Leigh—Deep and strong.

Langford—A good deep soil in the lower part of the parish.

Shipton—Sandy loam, pebbles, and stonebrash.

SECT. V.—WATER.

" In so far as the counties of Oxford and Berks are contiguous, they are separated from each other by the rivers Isis and Thames. The river Thame, which runs through the county, falls into the Isis at Dorchester, and from that place the river takes the name of Thames. Other rivers in Oxfordshire are, the Charwell, which divides this county from Northampton on a part of the boundary only; the Windrush, the Evenlode, the Glym, and the Ray; besides numerous streams of inferior note: so that this county may be considered as inferior to none, in point of being well watered."

There is a canal at Banbury, within thirteen miles

of

of Chipping Norton ; yet barley is carried in quantities by land, 44 miles, to Birmingham, and coals brought back, and sold as cheap as those from Banbury. This land-carriage system, a very miserable one for a good farmer, may be the fruit of no leases.

The Oxford Canal is marked in the annexed Plate.

CHAP.

CHAP. II.

STATE OF PROPERTY.

SECT. I.—ESTATES.

"IN regard to property, there are a few noblemen and gentlemen who have large estates, which, with the addition of the possessions belonging to the Church, and the different Corporate Bodies of the University, form a considerable portion of this county. There are also many proprietors of a middling size, and many small proprietors, particularly in the open fields."

In the red-land district, I found many considerable freeholders who farmed their own estates, and appeared to me to live in a respectable and comfortable manner, and many with the reputation of being wealthy.

I was informed, that there is in the county one estate, that produces above 20,000*l.* a year *on the table;* one of 12,000*l.*, one of 7000*l.*, one of 6000*l.*, one of 5500*l.*, two of 4000*l.*, and several of above 3000*l.*

SECT. II.—TENURES.

THE tenures by which land is held in Oxfordshire, are similar to those general through the South of England. Freehold and copyhold leases for lives remain
here ;

here ; and church and college leases, both for lives
and years, abound greatly. The fine usual, one year
and an half's rent ; but the rent itself raised in many
cases considerably.

SECT. III.—PRICE OF LAND,

May be estimated in Oxfordshire, in the opinion
of Mr. Turner, of Burford, at 26 years' purchase, at
a fair rent, yielding $3\frac{1}{2}$ per cent. on the capital invested.

Forest-hill, a farm of 372 acres, in the parish of the
same name, lett on lease of 21 years, five now (1807)
unexpired ; the rent, 497*l* , and valued lately at 800*l*.,
sold for 16,900*l*. ; estimated at 26 years' purchase.
The soil, sand and stonebrash arable, with good pasture and meadow.

Mr. Tormey purchased his estate at Burcot, two
years ago (243*l*. per annum freehold, and 131*l*. per
annum leasehold for three lives, all living, and all
under 30 years of age), for 13,000*l*. It contains a good
house, and a fishery in the river Thames : land-tax at
2*s*. in the pound on an old assessment, tithe-free.

CHAP. III.

BUILDINGS.

SECT. I.—HOUSES OF PROPRIETORS.

TO swell this article with descriptions of such a place as Blenheim, accounts of which are found in so many volumes, would be useless, and wide of the scope of such a Report. Suffice it to remark, that it will not be viewed without a lively interest, in its origin, its splendour, and the admirable spirit with which every thing is kept up to a becoming state of elegance, and the magnificence of the first residence possessed by a subject in Britain.

To such as would build a house and offices at the expense of about 20,000*l.*, I would recommend an attention to the plan of Mr. Wayland's house. It is not perhaps free from faults, but useful hints are to be taken from it.

SECT. II.—FARM-HOUSES AND OFFICES.

THERE is so much merit in one point of practice in Oxfordshire, that I hasten with pleasure to mention it : the capt stone reek-stands are well built, and general through the county ; and I have very rarely seen reek-yards better arranged, in point of regularity of position,

tion, proper spaces between the stacks, and neat husbandlike execution in building the stacks, and in the neatness of trimming and thatching them. They have a proper pride in a clean and well-ordered reek-yard, and are sure to walk a stranger into them. They form so perfect a contrast to the ragged heaps called stacks, by the courtesy of Suffolk and Norfolk, that I have returned to my own county and farm with no little disgust ; and must walk into five-and-twenty acres of well-dibbled wheat, in order to refresh an eye wearied with the deformities of the stacks.

Granaries built on capt stones are also very general, and well executed.

Mr. Cozins, of Golder, who occupies his own estate, has made considerable exertions, in nearly surrounding a large space with buildings, which he has divided into three yards; one for turning his teams into at night, another for his barren cows, and a third for the milch cows. He has built stables, cow-houses, and moveable barns ; and raised a new, and one of the best granaries I have seen, on capt stones ; it contains two stories, and the bins very conveniently disposed : the inside boarding is closely attached to the external, so as to exclude all harbour even for a mouse. It cost 180 guineas complete.

Another object well attended to in many cow and grazing farms, is the arrangement of the hay-stacks close behind the stalls, and feeding-houses, and yards, ready for the most convenient supply of the cattle through apertures in the wall, for the immediate delivery to the heads of the cattle. This is an important point in preventing waste, and saving labour.

But having noticed these particulars, the rest is a blank : I saw no truly well-contrived farmery, in which,

which due thought had been exerted in so placing
every building, that it could not possibly be moved
without incurring some inconvenience, greater than
arises from its present position.—Why is the stable
here?—Why are the cows there?—Why not have
placed the thrashing-mill in such a position?—The
answers to such questions shew, that essential objects
in these arrangements have not been in contemplation,
even in building new yards. The number of thrashing-
mills in the county is considerable; but all that I saw
have been placed with no other view, than that of finding a barn already built to cover it, and thus using a
building for the purpose, perhaps thrice as lofty as
was necessary. This is no cheap way of going to work;
and yet, with much space, there is generally no pro-
vision for the straw; and when chaff-cutters are annexed, none to receive the chaff, and still less attention
paid to the delivery of either to the animals that eat or
use them as food or litter. The position of a thrashing-
mill decides that of every other building; for it cuts,
or ought to cut, *all* the hay of the farm into chaff,
with much of the straw; and the house that immediately receives this chaff must be so placed, as to
admit a convenient delivery to the stables, stalls,
and sheep-yards. Thus the straw-house, chaff-house,
stables, stalls, hay-stacks, and sheep-yard, must be
placed in consequence of the position of the thrashing-mill, or waste and expense of labour must follow.
These, and many other circumstances that deserve attention, will be looked for in vain in the yards of Oxfordshire. Nor is it to be objected, that thrashing-
mills being a new invention, such attentions *could*
not be paid; for, in the first place, the new yards are
scarcely better, in this respect, than the old ones; and

in

in the next place, the same requisites are demanded under the old system of flail-thrashing, with no other difference than the thrashing-floor being the point of regulation instead of the mill, and a very small share of contrivance will adapt the system to more floors than one. But our views *at present*, since the discovery of the mill, demand quite a new arrangement.

SECT. III.—BUILDING MATERIALS AND LABOUR.

OXFORDSHIRE is under favourable circumstances in regard to a great plenty of building materials : stone and limestone abound in every part of the county ; slate for ordinary buildings is found in several districts; timber remains in parts of it to a tolerable amount ; but the new enclosures, as well as the open fields, are very naked : nor is the pay of artizans so high as in various parts of the kingdom.

Mr. Pratt, one of the stewards of His Grace the Duke of Marlborough, gave me the following prices : he has much knowlege in all that concerns timber, woods, and plantations.

1806.   Oak, 4*s.* to 5*s.* per foot, including top and bark ; elm, 1*s.* 3*d.* ditto; ash, 1*s.* 6*d.* to 2*s.*; beech, 1*s.* 3*d.*; sycamore, 1*s.* 6*d.*; lime, 1*s.* 3*d.* to 1*s.* 6*d.*

Some part sold to the Navy by those who buy the lots : the prices not so high as two years ago.

The Bishop of Durham, at Mungwell, has barn-floors made of beech, which, being cut down in winter, and when sawn out, laid in water for six weeks, have proved extremely durable.

" I do not think that the landlord would gain any thing

thing by paying *half* the workmanship : let him, his bailiff, or his steward, look over his farms occasionally, and the present mode of repairing will be found to answer as well as any plan of the kind ; but the best plan may perhaps be, for the landlord to pay the whole of the repairs, except straw and carriage, and to lett his farms accordingly. And as it is right that the tenant should have an interest in preserving the buildings, he should pay a certain sum per cent. upon all the workmanship in the necessary repairs. All buildings should be examined before the winter, and the principal workman should have orders to keep wind and water-tight till you see him again."—*Marquis of Blandford.*

SECT. IV.—COTTAGES.

THE exertions of the Board of Agriculture, soon after its establishment, and which have been continued to the present time, opened a new field for the benevolence of land proprietors to move in. The object of building comfortable cottages for the residence of the industrious labourer, and annexing a sufficient quantity of land to keep him from parochial support, became an exertion equally honourable to those who made it, and beneficial, in every view, to the poor people who were thus raised above a dependence on the rates of the parish. The advantages of the system were singularly experienced through two severe scarcities ; and as it was found, in a great variety of instances, that it completely stood the shock of those two most trying periods, it was pronounced by many persons of experience and ability, an object deserving of every attention

tion and encouragement. Inquiries into the state and condition of the poor became of more importance than ever, as well from the apparent utility of the plan recommended by the Board, as from the rapid and alarming rise of rates, where the old system has been adhered to.

A large party of intelligent farmers whom 1 met at Mr. Fane's, all agreed, that it was highly beneficial for cottagers to have each half an acre of land. Were it only for employment to keep them from the ale-house, the object is considerable, and would alone answer the expense; but it is advantageous in many other respects.

Formerly, many cottagers about Baldon had two, three, or four acres, and they kept cows; now, still having the land, they keep no cows: their rent, from 30*s.* to 42*s.* per acre, and all applied as arable. A query founded upon this—How far the system could do here? Probably the scarcities banished the cows: Did they not sell their cows, and plough the ground, tempted by the high price of wheat? But how came they permitted to plough? Did they not give up cows when butter was 20 per cent. cheaper than at present?

There are gardens, and good ones to nine-tenths of the cottages I have seen in Oxfordshire. A few years ago, they had no potatoes; now all have them. Formerly they did not like that root with their bacon, only cabbage: at present they are generally eaten. Sir Christopher Willoughby's cottages have not had their rents raised for near a century; but about Henley, &c. they pay from 3*l.* to 5*l.*

Mr. Turner, of Burford, has thirteen cottages, at the rent of 20*s.* to 45*s.*; all with gardens.

Cottages about Stonesfield, rent from 30*s.* to 3*l.* and
all

all have gardens. Labourers are plentiful in the county.

Many cottages at Bampton, &c. have gardens: rent, from nothing to 3*l.*; others pay 1*s.* per week.

The exertions of my Lord Bishop of Durham, in building cottages, are of uncommon merit, and almost equally deserving commendation, whether the design or the execution be considered. His Lordship has raised six pairs of them; they have each a very good garden, and conveniences for pigs. They cost building something above 100*l.* each, and two guineas is the rent paid for them. They are very substantially and conveniently built, as the annexed plan will shew.

The rent is so much below that of very inferior houses in the occupation of the poor, that it led to a conversation with the Rev. Mr. Durell, upon the system, which he explained with a propriety and a clearness not to be transfused by me in a cold recital: this will easily be believed by those who know the clearness of his conceptions and expression. Rent in money is not the object; but to place the inhabitants in such a state of ease and comfort as shall tend to habits of industry, sobriety, and honesty. Connected with, and uniting in, this object, as a part of the system, are other means strongly tending to the same end. Every workman employed by the Bishop is permitted to lodge, after harvest, in the bailiff's hands, 1*l.* 11*s.* 6*d.*; and in consequence of his readiness to make this reserve from a time of plenty, he is allowed through the following winter to purchase barley at 2*s.* a bushel under the market price, or any other product of the farm he may want, at a proportionable deduction in the price.

And further, to secure a benefit in buying such commodities

modities as the farm will not produce, his Lordship has established a village shop, at which they, and all the other poor of the neighbourhood, may buy whatever they want at a reduced price, for ready money. It was well remarked, that half the poverty of the poor arises from the various and mischievous evils of their running in debt to the village shopkeepers; evils that come with additional weight to the high prices they are usually forced to pay.

The wives and daughters of the cottagers receive all the flax they please, which is given them to spin into thread; and when they return such thread, they are paid the full price for spinning it. The Bishop has it woven into cloth, according to the fineness of the spinning; and this is sold to the cottagers' families at 2d. a yard lower than the ordinary price.

Combining these advantages of very light rents, good gardens, plenty of pigs, stocks of bees (one originally given to each cottage), the farm products commanded at an easy price, and all others had at a low rate, without debts—these families are placed in a situation productive of good morals. None of them ever receive any relief from the parish.

Should it be observed that all this is well for a man of great fortune, but inapplicable to others—it must be remarked, that it is applicable to all farming their own land, and able to build cottages: the benefit of the measure to the employer here becomes an object of consequence. Setting aside any attention to personal merit, and benevolence, and feelings which are not universal, the system has advantages, perhaps adequate, on a different calculation, to prove that it is profitable as well as benevolent. To employ no other labourers on a farm than such as are placed above the tempta-

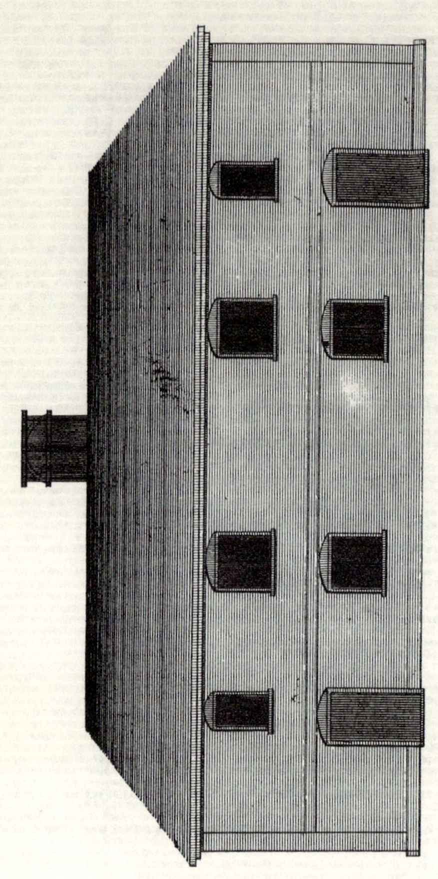
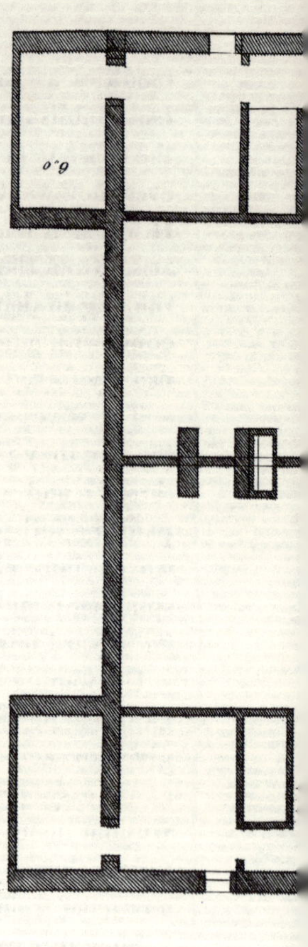

The Bishop of Durham's Cottages.

Pig Sty
Celler
the cellor runs under the stairs
10.6
10.6
15.0
15.0
10.6
15.6
Nede sc. Strand.
10 20 30 40 50 60 70 80

temptations resulting from an abject poverty, is alone a point as clearly beneficial in the general idea, as it is unsusceptible of a minute money calculation. He who has farmed long and largely, must know the losses incident to ill neighbourhood: barns are scenes of plunder, and fences and woods of depredation; vigilance in the preservation of rural property, is as necessary as industry to acquire it; and every magistrate has cause to know, that with all the watchfulness and anxiety of farmers, their losses are not inconsiderable. By such a system as the Bishop's, you go to the root of the evil, in the poverty and disadvantages of the labouring poor, and gradually form them to habits which mature the effect, and render it durable. This is, and must be, as profitable to the employer as it is to the labourer.

I know but one grand objection to be made, and that is—*new* and *fashionable*: men and women thus placed in a state of ease and comfort, will breed children most intolerably; and as the increase of population demands *checks* rather than encouragement, the Bishop has quite mistaken the matter, and is doing the public an injury, while he most erroneously thinks himself well employed. This we leave to the politicians, who, if they agree in the opinion, have feelings not much to be envied. One remark I shall venture, though a political one: the regularly acting check to population in this kingdom, is not the price of corn, nor any thing connected with it, but the difficulty of obtaining habitations. This operates at all times, and in all places. It is the only reason why marriages do not multiply; it is a check entirely in the hands of those who have the means of measuring both the want and the superfluity. In the application it is, as it ought to be,

local;

local : you have the power to pull down and to build up, and you can do neither without the effect surely following the choice dictated by the circumstances of the district. Go to a place where a manufacture is permanently established, and there it may not be wise to add to cottages—it may be *prudent* even to pull down, when opportunity offers: go to the fens of Lincolnshire, and you will soon see the propriety of building many new ones. But you can go no where, where it is not practically and politically wise, to change wretched hovels into comfortable dwellings, and so to arrange the adjacent land, as to take families from a dependence on the parish rates, and enable them, by industry and sobriety, to maintain themselves.

" It has been recommended, on former occasions, to lay to the cottages of labourers a small portion of land, in such places as might be converted into pasture ; and though I must for the present produce the two following instances of the effect of such a proceeding from two neighbouring counties, yet the instances appear to me to be too much in point to be here omitted. It is almost needless to say, the authenticity of them is unquestionable.

" The commonable land belonging to a parish in Worcestershire, which is situated very near to Tewkesbury, in Gloucestershire, was enclosed about 22 years ago ; and there was an allotment, containing 25 acres, set out for the use of such of the poor as rented less than 10*l.* a year, to be stocked in common. At that time there were about sixteen people on the parish books, some of whom had families. Previous to the enclosure, there were some few cottages that had land lett with them, to the amount of 6*l.* or 7*l.* a year each. The

occupiers

occupiers of these cottages with land annexed to them, were remarkable for bringing up their families in a more neat and decent manner than those whose cottages were without land; and it was this circumstance which induced the lord of the manor (to whom almost the whole of the parish belonged) to lay a plot of land, from five to twelve acres (besides the common before-mentioned), to other of the cottages, and to add a small building, sufficient to contain a horse or a cow, and likewise to allow grafting stocks to raise orchards. In some instances, small sums of money were lent to these cottagers, for the purchase of a cow, a mare, or a pig.

" The following good effects have been the consequence of this proceeding: it has not, in one instance, failed of giving an industrious turn, even to some who were before idle and profligate; their attention in nursing up the young trees has been so much beyond what a farmer, intent upon greater objects, can or will bestow, that the value of the orchards is increased to 40s. per acre, in land which was of less than half the value in its former state; and the poor's-rates have, from this cause, fallen to 4d. in the pound, or less, there being only two (and those very old) people on the books at this time, whilst the adjoining parishes are assessed from 2s. 6d. to 5s. in the pound. These are labourers, and good ones: their little concerns are managed by their wives and children, with their own assistance after their day's work. Their stock consists of a cow, a yearling heifer, or a mare, to breed (from which a colt at half a year old will fetch from 3l. to 5l.), a sow, and 30 or 40 geese. This, therefore, has been the means of bringing a supply of poultry and fruit to the market, of increasing population, and making the land produce double the rent a farmer can afford to give.

" The

" The other parish to which I alluded, is situated within six miles of Leicester, and belongs to a nobleman whose family have for many years lett small quantities of land, varying from four to twenty acres, with the cottages, after the rate of about one-fifth less than the same sort is lett for to the farmers. These cottages keep from one to three or four milch cows to make butter, from five to twenty lambs (being chiefly twins, purchased, and brought up by hand), one or more pigs, and raise from one to three or four young beasts yearly.

" The consequence is, that about twenty families live comfortable as labourers, whilst the management of their stock employs their families, and themselves at their leisure time, which might otherwise be spent in an ale-house. The poor's-rates are only from 6*d.* to 8*d.* in the pound, which may be considered as a saving to the parish of 60*l.* or 80*l.* a year. It is true, that the landlord sustains a loss, in the first instance, of about 80*l.* a year in rent, on account of these lands being lett cheaper than the farms; but it is doubly restored, by enabling the farmers to pay a greater rent for their farms, on account of the poor's-rates being so easy*."

* Mr. Davis, original Report.

CHAP. IV.

OCCUPATION.

SECT. I.—FARMS.—FARMERS.

" THE size of the farms varies so much, that it is difficult to speak on that head ; but they may be considered, generally speaking, as less than in most parts of England*."

The more that inquiries are multiplied in the kingdom, the more clearly it will be shewn, that farms are not too large: supposing that one or two in a county may be found that might be advantageously divided, such exceptions are so rare, that to found any general reasoning upon them would be futile. Upon the whole, they are too small to be consistent with good husbandry, and I found this to be the fact in many parts of Oxfordshire.

About Stoken-Ash, none larger than for 200 or 300 sheep.

The largest farms in the rich Thame district, do not usually exceed 300 acres.

For some miles around Blenheim, farms are in general from 100 to 500 acres : here and there a larger, and some smaller. The great farmers are generally the best cultivators.

---

* Original Report.

Farms

Farms at Dorchester.—A rate at 1*s.* in the pound, raises 96*l.* 19*s.*

No. 1.		£.22	3	0
2.		15	17	0
3.		12	4	0
4.		7	3	0
5.		6	0	0
6.		5	18	0
7.		4	9	0
8.		4	8	0
9.		2	17	0
10.		2	0	0
11.		1	15	0
12.		1	6	0
13.		1	0	0
14.		0	18	0
15.		0	14	0
16.		0	10	0
17.		0	8	0
18.		0	7	0
19.		0	7	0
20.		0	7	0
21.		0	7	0
22.		0	8	0
23.		0	8	0
24.		0	8	0
25.		0	6	0
26.		0	6	0
27.		0	6	0
28.		0	6	0
29.		0	6	0

Carry forward, ... £.93 12 0

Brought

		£.	s.	d.
Brought forward,		93	12	0
No. 30.	....	0	6	0
31.	....	0	5	0
32.	....	0	4	0
33.	....	0	4	0
34.	....	0	4	0
35.	....	0	4	0
36.	....	0	4	0
37.	....	0	3	0
38.	....	0	3	0
39.	....	0	3	0
40.	....	0	3	0
41.	....	0	2	0
42.	....	0	2	0
43.	....	0	2	0
44.	....	0	2	0
45.	....	0	2	0
46.	....	0	2	0
47.	....	0	2	0
48.	....	0	2	0
49.	....	0	2	0
50.	....	0	2	0
51.	....	0	2	0
52.	....	0	2	0
		£. 96	19	0

Over Norton and Salford.—One of 600*l.* a year, one of 400*l.*, one of 500*l.*, one of 200*l.*, one of 100*l.*, one of 450*l.*

On the very fine estate of G. Stratton, Esq. at Great Tew, there are the following :

No. 1.

No. 1.	....	£.16	16	0
2.	....	210	0	0
3.	....	172	0	0
4.	....	302	0	0
5.	....	75	0	0
6.	....	240	0	0
7.	....	243	0	0
8.	....	110	0	0
9.	....	220	0	0
10.	....	560	0	0
11.	....	210	0	0
12.	....	190	0	0
13.	....	360	0	0
14.	....	343	0	0
15.	....	225	0	0
16.	....	10	0	0
17.	....	6	6	0
18.	....	59	14	3
19.	....	12	7	6
20.	....	255	0	0
55 cottages, each with a very good garden		126	19	0
		£.3943	2	9

Besides a house and gardens, &c. 64 acres of very fine and thriving plantations.

For ten miles around Baldon, there are no larger farms than from four to 500*l.* a year, and very few above 250*l.* or 300*l.*

The design of Sir C. Willoughby's farming is to raise every object of the consumption of a family of 30, at home, that the climate of the country will give. He annually kills 80 sheep; by agreement with a neighbour, he eats his own beef. He keeps nineteen cows for butter, milk, cream, and cheese. A productive dove-house yields an ample supply of pigeons. His ponds (having a small stream through them, and being well attended) afford him carp, tench, and perch : carp of three to six pounds, tench one pound, perch from a half to two pounds, and to be had whenever he-wants them. Poultry of all sorts in great abundance. Game. His own wheat, oats, and hay; makes his own malt, and raises hops and poles. All this forms a system of family plenty, and gives the satisfaction of every thing being good of the sort, if due attention be paid to the management; and when it is effected by little other expense than the labour and taxes of a farm of less than 400 acres, for the supply of so large a family, with a considerable surplus of many articles for sale, it is, in the mind of this very attentive and reflecting proprietor, a proof that the system is not only pleasant, but profitable ; and of this fact I have no doubt, with so large a family.

The *times* give no inconsiderable importance to notes of this complexion; taxes, and the rise in the price of so many articles of a gentleman's consumption; but above all, the modes of living are become such, as to threaten the comforts of many classes who, twenty or thirty years ago, thought themselves moving in a sphere above all apprehension. The *nearest way of going to work*, is now well worth attending to.

" There are many farms in the Forest district, from 20*l.* to 80*l.* per ann.*"

" In some counties large farms are much more productive in the gross than small ones : they may not

---

* Marquis of Blandford's Annotations on the Original Report.

produce

produce so many chickens and ducks, but certainly more mutton and wool, and more beef and corn, and what some people think of more consequence, more children, by employing a much greater number of labourers."—*J. Boys*.

FARMERS.

In regard to the character of the Oxfordshire farmers a remark may be made at present, which will not probably be just twenty years hence; and I well know, was not the case twenty, thirty, and forty years ago: when I found them to be a very different race from what they are at present. They are now in the period of a great change in their ideas, knowledge, practice, and other circumstances. Enclosing to a greater proportional amount than in almost any other county in the kingdom, has changed the men as much as it has improved the country: they are now in the ebullition of this change: a vast amelioration has been wrought, and is working; and a great deal of ignorance and barbarity remains. The Goths and Vandals of open fields touch the civilization of enclosures. Men have been taught to think, and till that moment arrives, nothing can be done effectively. When I passed from the conversation of the farmers I was recommended to call on, to that of men whom chance threw in my way, I seemed to have lost a century in time, or to have moved 1000 miles in a day. Liberal communication, the result of enlarged ideas, was contrasted with a dark ignorance under the covert of wise suspicions; a sullen reserve lest landlords should be rendered too knowing; and false information given under the hope that it might deceive: were in such opposition, that it was easy to see *the change*, however it might work, had

not

not done its business. The old open-field school must die off before new ideas can become generally rooted.

SECT. II.—RENT.

I FOUND among the farmers throughout the county, a jealousy relative to this subject, which, since the property-tax, must be considered as ill exerted Government have attained, to a degree of certainty, much more knowledge on this subject than was ever possessed before; and in every county the means of knowing it are so increased, that concealment is idle, so far as agricultural inquiries can be concerned. But the Board orders such an article to stand in the inquiries of Surveyors, not for any political purpose, but with the mere view of better understanding the various descriptions given of the soils. Light and heavy, weak and strong, are such indefinite terms, that any addition, like that of rent or produce, become an explanation in doubtful cases. In using such terms, those and many more might be equally applied to some poorer sands in Dorchester hundred, and to the fertile ones of Atterbury; yet these are worth 40s. per acre better than the first merit 15s.

The upper part of Mungwell parish was within ten years, 10s. an acre. At present the whole would lett for 25s. Rent of the Thames meadows about 3l. They are better towards Reading.

About English, 17s. 6d.

At Whitchurch, some pay for the vale 45s. per acre round. The hills 20s.

From various notes through the Chiltern district, it
appears

appears that the rents rise from 12*s*. to 30*s*.; and some land on the Thames 40*s*. Were I to estimate the average of the whole district as marked in the map, I should name 16*s*. as the average rent.

About the Baldons, &c. open-field arable twenty years ago 7*s*. 6*d*. per acre; now 16*s*. and the farmers much richer now than they were then.

In the last two or three years rents are raised greatly in all the vicinity of the Baldons; but not near doubled. Most inconvertible land. College fines at reserved rents, on 21 years leases, one year and a half.

The open-field at Dorchester, letts at 25*s*. per acre. It is red sand.

Sir Gregory Turner letts his property all round, at Merton, at 3*l*. per acre: Exeter College lett theirs at the same place at 40*s*.

Mr. Rowland, at Water Eaton, on land all round at 63*s*.: keeps many and great beasts.

Hampton Poyle enclosed in 1797: rents all round, meadow and arable, at from 35*s*. to 40*s*. per acre.

In the rich district of Thames and North Weston, a meadow of 50 acres was lett many years ago for 100 guineas. Some part of the district letts at 3*l*. and many whole farms throughout at 45*s*.

Wendlebury, before the enclosure about seven years ago, was 9*s*. or 10*s*. per acre. The Commissioners' quality price was from 13*s*. to 40*s*. and the average 24*s*. tithe-free. The quantity 1100 acres.

At Black Bourton, there was a valuation at 18*d*. or 2*s*. per acre, 35 years ago, which now is 14*s*.

Bampton from 12*s*. to 15*s*. the field acre. An enclosure here would treble the Earl of Coventry's rents; perhaps quadruple them. They have a common of

2000

2000 acres of *green* sward, which enclosed would be *famous* land.

It is extremely difficult to estimate so various a district as that comprehended under the denomination *miscellaneous*: perhaps 25*s*. may not be far from the fact.

I was assured at Bloxham, that a farm of 350 acres there would lett at 50*s*. per acre round; but the land is not so good as the red sand at Adderbury.

At and about Great Tew, arable 15*s*. to 26*s*.; grass 28*s*. to 40*s*.; rates 3*s*. 6*d*. in the pound.

I am not beyond the truth, in conjecturing that the red land district letts or is occupied at the rent of 30*s*. on an average.

Average of the stonebrash around Coversfield, estimated at 20*s*.

Westcot Barton, stonebrash 20*s*. per acre, tithe-free.

Average of the county, 25*s*.

Duns Tew 24*s*.; was before enclosing 12*s*. or 13*s*.

In general, in the whole stonebrash district, rents are doubled by enclosure; in some cases trebled.

I estimate the whole stonebrash district at 20*s*. per acre.

In the district south of Oxford, I heard the rent of the whole county estimated at 20*s*. all round on an average, wastes included.

A very considerable person thought that this was too high, and that it did not amount to above 17*s*.

A third, reckoned all profitable land at an average of 24*s*.

Acres

Red land,	79,635	at 30s.	£.119,452	10
Stonebrash,	164,023	at 20s.	164,023	0
Chiltern,	64,778	at 16s.	51,822	8
Miscellaneous,	166,400	at 25s.	208,000	0
Total	474,836	Ave. 22s. 10d.	£.543,297	18

SECT. III.—TITHE.

Wotton, enclosed by Act of Parliament, and yet left subject to tithes.

Mr. Turner letts 2000 acres of tithe at 4s. an acre all round, for corn tithe, on stonebrash land.

At English, &c. 5s. in the pound; at Golder 5s. per acre; for grass 2s. 6d. to 3s.; Crowmarsh 6s. to 7s. per acre below hill; but less for the hills.

In general, of arable land fairly lett, one-fourth of the rent. The University of Oxford is supposed to possess the property in whole or in part, of one-sixth of the county. By the expression *in part,* I conceive tithes to be meant.

" It would be extremely desirable to institute an inquiry into the comparative state of cultivation of the land held under colleges or ecclesiastics; the farm buildings, and the circumstances of the tenants, that some conclusion might be drawn on the advantages or disadvantages of different tenures. It might be incidental to such inquiry, to estimate the amount of capital required to support a farm sufficient for the maintenance of a family at a rack-rent, and under college tenure; as also the size of a farm under college tenure and at rack-rent, which, with the same capital, would support a family. My observations tend to shew that the

the same capital would better support a family on a farm of the same value at rack-rent than under college tenure.   Rack-rent being payable half-yearly, requires not the foresight that is required to provide for a fine payable at the renewal of a life, or at the end of seven years; it is more congenial to a farmer's prudential practice, which seldom reaches beyond the spur of the occasion.   When there is a distinct fortune, independent of the farm capital, college tenure proceeds smoothly; the buildings look respectable, and the tenant lives in comfort; but in other leases the farm and the farmers gradually fall into decay."—*MS. Annot.*

In general, one-third of the rental, or one-fourth, tithe included; one-fifth of the rent on poor land, and 1*s.* 6*d.* over per acre.   This, if under a pound, from 10*d.* to 20*d.*

In most of the enclosures that have taken place about Bicester, one-ninth and one-fifth have been given for tithe.   At Lower Heyford one-fifth and one-ninth. At Hampton Poyle the same, but the land not assigned; a corn-rent.   The Commissioners fixed the bushels payable by each person, according to the quality of the land, and the price is regulated every year on that of Oxford market at Lady-day and Michaelmas.

Tithe at Bampton taken in kind; when compounded, 3*s.* per acre arable, and 2*s.* grass.   They were offered for 3*s.* 8*d.* and refused by the farmers, who were surely no conjurers, for 4*s.* were given to gather.

Tithe is various in Oxfordshire; a few rectors have one in fifteen, and others one in twenty.   This arose from varied endowments; half the tithe was settled on the rector, and the other half perhaps given to some religious house, and on the suppression came into lay hands.

hands. That the tithes came from the land-owner, appears from the manor farm, and that only, being in so many cases tithe-free.

" Corn-rent in lieu of tithes, is perhaps the most generally approved method of commutation, and yet, when fully examined, it will be found one of the most absurd.

" The allotment of land seems the most fair and equitable of all others; and yet neither this nor any other could prove an equivalent. The average here alluded to might be advanced from 105*l.* to 220*l.*; but what would have been the real value of the tithes after the enclosures? Without this information the rent is of little consequence. The impossibility of an equivalent for tithes, however, though apparently a powerful objection to a commutation, is in reality the most irresistible argument in its favour.

" Whether manure fertilizes private property, or parochial glebe, is indeed a matter of no importance to the public, provided there be the same produce in the whole parish. But this, it might easily be shewn, is in many instances impossible, and in most, highly improbable. An additional objection to tithes in kind, and perhaps a more powerful one than any here stated, is, that the gathering them in will be generally 20*l.* per cent. sometimes 30*l.* greater to the tithe-owner than to the farmer. But the most irresistible objection to tithes, whether taken or not taken in kind is, that in real value they increase in a most disproportionate degree to the increase of the value of land."—*Rev. John Howlett.*

Commutation.—" It has been disputed what is the best

best system to follow when tithes are to be commuted. In this county, many enclosures have taken place within a few years, wherein all the several methods have been pursued. In divers of these enclosures the land has been left titheable as before, because the tithe-owner and proprietors did not agree on terms. In others, an annual rent has been fixed, to be paid out of each estate, varying according to the prices of corn, taken at stated times ; and this method has been satisfactory in many cases. But the most usual mode is to set out an allotment of land in lieu of tithes, by which both rectorial and vicarial estates are often greatly improved in value : amongst other instances, I am favoured with the particulars (too copious to give here in detail) of a vicarage near Banbury, which improved from 105*l.* to 220*l.* a year immediately upon the enclosure; and at the expiration of a twenty-one years lease, the value was further considerably increased.

" On the subject of commutation of tithes, however, if the matter is fairly viewed, it is right, briefly at least, to observe what is said on the other side of the question.

" It is undeniable that, as matters are at present, agriculture is daily improving, and therefore, though it must be confessed a desirable object to exonerate lands from tithes, yet it may be doubted whether they are so great an obstacle to improvements as sometimes represented. If a farmer occupies land of two descriptions, one portion titheable, the other exempt from tithes, it is natural to suppose he will be more anxious to manure that which will return him the entire profits, than that from which he is to receive a part only of the produce. He pays a specific sum for

his

his farm; but from what part or parcel of land the money accrues is indifferent. His attention will be directed to that which, in the least given compass, and with the least expense, will render the largest profits. But cases of this sort, comparatively rare and few in number, are not the proper instances to argue upon.

" As to the objection of carrying the profits (when the tithes are taken in kind) to other lands, it is obviated, if the manure so made is properly applied. The profits arising in a parish, are expended in the parish; and whether they fertilize private property, or parochial glebe, the general produce is equally improved, and the public equally benefited.

" Of the force of these arguments I hazard no opinion; but I should think myself deficient in the discharge of the commission confided to me, if, together with the most interesting facts, I did not also state the most material observations which I have heard, or which have occurred to me. That Honourable Board, to whose consideration this Report is with all deference submitted, will bestow on the particulars that degree of attention which they shall respectively appear to deserve."—*Original Report.*

SECT. IV.——RATES.

Mr. Sotham estimates poor-rates in Oxfordshire, on an average, at 4*s.* to 4*s.* 6*d.* in the pound.

Mr. Pratt conceives them to be on an average, in the parishes with which he is acquainted (which are many) from 4*s.* to 6*s.* in the pound.

At Bensington 8*s.* in the pound; were 14*s.* in the scarcity.

Never

Never less at Thame than 6s. in the pound.

At Golder, 5s. in the pound; Crowmarsh, 6s. to 7s. in the pound; Mungwell, 4s. in the pound; English, 4s. in the pound; Whitechurch, 8s. in the pound.

At Bampton 5s. in the pound.

Kempcott, rates 2s. and tithe-free.

At Burford, 10s. and 11s. in the pound; in country parishes 3s. to 5s. in the pound.

At Witney, 10s. in the pound; at Ramsden they were once 50s. in the pound.

General average 3s. 4d. Bloxham, &c. 5s.; at Bloxham 70 people on the round : many shag weavers out of employment.   Many such weavers in Banbury, &c.; a rough coarse velvet made there : perhaps 1000 have been employed; but not at present: bad for farmers, ready for the parish, and a heavy burthen to the country; nor so well disposed as the country labourers. This fabric, like that of Witney, has travelled to the North.   The unemployed, sent round from farmer to farmer for employment, were beginning to riot, had not the yeomanry come in.

At Dorchester, 7s. or 8s. in the pound.

Disbursements.

December 1771, to April 1772 ....£.49   0   0
April 1772, to November 65   8 11
November 1772, to April 1773 63   5   3
April to October 1800 .... 453 18   6
October 1800, to April 1801 .... 671 19 11
April to October 1801 480   0   4
October 1801, to April 1802 393   1 10
April to October 1802 .... 354 19   0
October 1802, to April 1803 314   7   9
April to October 1803 .... 236 10   0

October

October 1803, to April 1804 .... £.292 8 10
April to October 1804 .... 284 10 9
November 1804, to April 1805 .... 364 8 4
May to October 1805 .... 300 6 8
October 1805, to April 1806 .... 361 9 10
April to October 1806 .... 388 18 2
October 1806, to March 1807 .... 411 12 6

At Burcot, no rate was levied, and poor people were got to take collection, in order to prevent the parish being taxed in aid of Dorchester. At Brightwalton, in the scarcity 29*s.* in the pound.

At Chimner, 7*s.* to 9*s.* in the pound; in the scarcity they were 12*s.* to 14*s.* At Crowell 6*s.*; Lower Heyford 3*s.*

In general, in the parishes about Caversfield, Mr. Bullock estimates them at 3*s.* to 4*s.* in the pound.

The county-rates in Oxfordshire, by the extraordinary good management of the magistrates, and the highly to be commended exertions of the gentleman who has so ably filled the Chair of the Sessions for above twenty years, have been kept down in a manner that should be an example to other counties. It does not amount, on an average, even to 3*d.* in the pound; and yet a new gaol has been built at Oxford: of the management of which, details have been spread through the kingdom, and I believe have been productive of much good in various districts.

It appeared by the returns to Parliament in 1803, that the average of the county was 4*s.* 8*d.*

Summary

*Summary of the County Totals of the Returns rela-
tive to the Expense and Maintenance of the Poor,
for the Year ending Easter, 1803; to which is sub-
joined, the Amount of Charitable Donations.*

Total money raised by the poor's-
rate and other rate or rates,
within the year ending Easter,
1803 .... £.103,559 10 6

Amount of similar rates; return
made under 26 Géo. III. c. 56,
1786; medium average of 1783,
1784, and 1785 ....   38,348 8 11

Amount of similar rates, the return
made under 16 Geo. III. c. 40,
1776, money raised in the year
ending Easter 1776 ....   31,154 12 7

At what rate in the pound for the
year ending Easter, 1803 ....   0 4 8

Expenditure on account of the poor,
1803 ....   91,304 9 8

Militia, highways, church, county,
&c. 1803 ....   11,836 19 7

Earned by the poor out of houses 1221 4 0
Ditto in any house 1270 15 8

Number of persons relieved from
the poor-rate permanently, *out*
of houses, not including children   6539

Ditto *in* houses   1243

Number of children of persons relieved permanently, *out* of houses,
and under five years of age   3214

Ditto from five to fourteen years   4841

                                                      Number

Number of persons relieved occa-sionally	6148
Number of persons included in the three preceding articles, above 60 years of age, or disabled from labour by permanent illness or other infirmity	2912
Number of persons relieved, not be-ing parishioners	2800
Number of friendly societies	69
Number of members in the said so-cieties	5010
Number of children in schools of industry	290
Amount of charitable donations, 1786, annual produce of money	£.706 13 3
Ditto from land	3347 2 11
Total annual produce	£.4053 16 2

POOR-RATES OF OXFORDSHIRE.

Bampton Hundred.

Name and Description of each Parish or Place.	Total Money raised by the Poor's-rates, and other Rate or Rates, within the Year ending Easter, 1803.				At what Rate in the Pound, for the Year ending Easter, 1803.		
Alvescott	£.357	18	1	....	£.0	3	11
Asthall, and Asthall-leigh	208	4	10¼	....	0	3	7
Aston and Coat	672	11	8	....	0	5	8
Bampton and Weald	995	7	4½	....	0	6	0
Black Bourton	398	13	7	....	0	5	2
Brighthampton	54	0	0	....	0	10	6
Broadwell	208	11	9¼	....	0	3	0
Broughton Poggs	259	16	4½	....	0	7	3½
Burford	1300	7	3¾	....	0	15	0
Chimney	36	7	9	....	0	1	9
Clanfield	289	11	0	....	0	4	0
Crawley	320	7	2½	...	0	9	0
Curbridge	520	16	8	....	0	15	0
Ducklington	357	9	2	....	0	10	0
Filkins	655	0	0	....	0	9	5
Grafton	49	6	2	....	0	2	9½
Hailey (F)	1142	4	4	....	0	10	9
Hardwicke and part of Yelford }	29	19	8	....	0	4	0
Holwell	70	2	6	....	0	2	9
Kelmscott	118	14	2	....	0	2	0
Kencott	208	1	11¼	....	0	5	3
Lew	232	13	3	....	0	6	4½
Norton, Brize	463	5	10½	....	0	1	6
Radcutt	24	3	2	....	0	0	10½
Shifford	31	9	2	....	0	1	0
Standlake	334	11	9¾	....	0	10	6
Upton and Signet	390	9	0	....	0	3	4
Westwell	123	12	0	....	0	3	0
Witney	1887	3	4½	....	0	13	6
	£.11,740	19	2¾	Av.	£.0	6	1¼

Banbury

Banbury Hundred.

Name and Description of each Parish or Place.	Total Money raised by the Poor's-rates, and other Rate or Rates, within the Year ending Easter, 1803.			At what Rate in the Pound, for the Year ending Easter, 1803.		
Banbury	£.2019	2	6	£.1	12	6
Bourton, Great and Little	551	8	0	0	5	9
Charlbury	553	4	8¾	0	11	6
Clattercott	28	8	11½	0	1	5
Claydon	269	6	6½	0	2	10½
Cropedy	544	4	9	0	5	3
Epwell	293	13	0½		—	
Fawler	143	1	8	0	10	0
Finstock	290	3	11	0	7	6
Neithrop (F)	1377	19	5½	0	9	0
Prescott	52	10	6	0	1	6
Shutford, East	38	14	4½	0	0	10½
Shutford, West	264	3	7	0	5	0
Swalcliffe	361	1	7¼	0	5	8
Wardington (F)	1148	2	6	0	9	11
	£.7935	6	1½	Av. £.0	7	9¼

Binfield Hundred.

Bix	£.420	15	5.	£.0	5	8
Caversham	746	16	11	0	4	0
Eye and Dunsden	616	7	9	0	5	9
Harpsden (with Bolney)	367	6	6	0	7	7½
Henley-upon-Thames ..	2603	16	5¾	0	6	9
Rothersfield Grays	779	1	0	0	6	0
Rotherfield Peppard	367	12	3	0	5	6
Shiplake	458	3	3	0	5	0
	£.6359	19	6¾	Av. £.0	5	2½

Bloxham Hundred.

Name and Description of each Parish or Place.	Total Money raised by the Poor's-rates, and other Rate or Rates, within the Year ending Easter, 1803.	At what Rate in the Pound, for the Year ending Easter, 1803.
Adderbury, East	£.689 16 4½	.... £.0 4 11½
Adderbury, West	370 11 8½	.... 0 0 6
Alkerton	54 10 0	.... 0 2 0
Barford, St John's	104 15 0½	.... 0 2 5
Bloxham	1109 7 8½	.... 0 5 0
Bodicot ....,.................	427 0 10¾	.... 0 6 0½
Broughton	254 18 3	.... 0 3 4½
Drayton ....,.................	211 18 9	.... 0 2 6
Hanwell	166 13 8½	.... 0 3 1
Horley	380 5 9	.... 0 6 0
Hornton	343 0 3¾	.... 0 7 6
Milcombe	309 7 5½	.... 0 4 9¼
Milton	187 5 1¼	.... 0 4 2
Mollington	258 15 11	.... 0 4 9
Newington, North	286 1 10¾	.... 0 5 6
Sibberd Ferris	243 14 11	.... 0 5 2¼
Sibford Gower	472 4 7¾	.... 0 5 0
Tadmarton	355 0 2	.... 0 3 8
Wiggenton	259 10 9	.... 0 4 4½
Wroxton	506 16 5	.... 0 5 3

£.6991 15 9¼   Av. £.0 4 3½

Bullington Hundred.

Name and Description of each Parish or Place.	Total Money raised by the Poor's-rates, and other Rate or Rates, within the Year ending Easter, 1803.				At what Rate in the Pound, for the Year ending Easter, 1803.		
Albury	£.28	10	4	.... £	0	1	0
Ambrosden	67	8	0	....	0	2	10½
Arncott	260	4	9½	....	0	4	5
Arncott	210	6	1½	....	0	4	5
Baldon, Marsh	156	0	7	....	0	5	3
Baldon, Toot	219	9	0	....	0	2	2
Beckley	283	13	0	....	0	6	0
Blackthorne	349	8	11¾	....	0	5	6½
Chilworth	87	8	6	....	0	1	3
Chippinghurst	28	0	0	....		—	
Clements, St.	212	6	2	....	0	5	5¼
Cowleys	135	17	4	....	0	0	11
Cuddesden	195	10	3	....	0	2	6
Denton	120	0	0	....	0	3	0
Eldsfield	92	2	6	....	0	2	6
Forest, Hill	78	0	0	....		—	
Garsington	470	17	9¾	....	0	5	4½
Headington	434	4	7	....	0	4	3
Holton	129	19	3	....	0	1	5¼
Horsepath	137	4	9½	....	0	1	3
Iffley	352	6	9½	....	0	1	11½
Littlemore	15	8	4	....		—	
Marston	351	0	0	....	0	6	0
Merton	219	4	9½	....	0	1	0
Nuneham Courtney	241	3	0	....	0	2	3½
Piddington	448	1	0	....	0	5	2
Sandford	175	18	5¼	....	0	1	11
Shotover	55	16	0	....	0	2	0
Stanton St. John	355	13	1	....	0	3	5
Stow-wood	37	8	6	....	0	3	0
Studley and Horton	186	10	1	....	0	5	8
Tiddington	31	11	8	....	0	1	6
Waterperry	173	19	8	....	0	2	3
Wheatley	434	15	3½	....	0	7	0
Wood Eaton	64	14	9	....	0	2	0
	£.6840	3	4¾	Av. £.0	3	1	

Chadling-

Chadlington Hundred.

Name and Description of each Parish or Place.	Total Money raised by the Poor's-rates, and other Rate or Rates, within the Year ending Easter, 1803.			At what Rate in the Pound, for the Year ending Easter, 1803.		
Ascott	£. 381	13	6	£. 0	8	0
Bruern	24	4	6	—		
Chadlington, East and West	364	3	5	—		
Chastletown	243	3	0	0	3	0
Churchill	451	5	0	0	3	5½
Cornwell	122	4	3	0	2	7
Enston	740	3	10¼	—		
Fifield	94	1	1½	0	3	1½
Fulbrook	315	0	0	0	4	0
Heythrope (with Dunthrop)	158	3	1	0	0	6
Idbury	192	6	8	0	2	0
Kingham	246	14	0	0	4	0
Langley	38	11	0	0	1	0
Leafield	305	12	0	0	9	0
Lyneham	265	7	6	0	3	11
Milton	356	4	0	0	2	1
Minster Lovell (F)	489	5	7	0	7	0
Northmoor	369	15	0	0	5	10
Norton, Chipping	1201	5	5	0	6	3
Norton, Hook	1203	11	8	0	6	6½
Norton, Over	648	17	0	0	7	0
Ramsdon	363	8	0	0	1	3
Rollright, Great	557	13	10	0	4	6
Rollright, Little	70	0	0	0	3	6
Salford	229	9	5½	0	4	0

Name and Description of each Parish or Place.	Total Money raised by the Poor's-rates, and other Rate or Rates, within the Year ending Easter, 1803.	At what Rate in the Pound, for the Year ending Easter, 1803.
Sarsden	£.196 12 6½	....£.0 3 9
Shipton-under-Which-wood	562 2 11	.... 0 1 0
Shorthampton	194 18 10½	.... 0 3 0
Spelsbury	473 15 5	.... —
Swerford	203 14 8½	.... 0 3 5
Swinbrook	85 14 0	.... 0 4 0
Taynton	242 4 10	.... 0 6 2½
	£.11,391 6 1¾	Av.£.0 4 10

Dorchester Hundred.

Name and Description of each Parish or Place.	Total Money raised by the Poor's-rates.	At what Rate in the Pound.
Burcott	£.76 15 9	£.0 2 8
Chistlehampton	139 19 7½	.... 0 1 10
Cliften-hampden	88 0 6	.... 0 1 9
Culham	411 17 8¼	.... 0 1 1
Dorchester	675 18 0	.... 0 7 2
Drayton	219 18 4½	.... 0 3 9½
Southstoke	670 9 0	.... 0 8 9
Stadhampton	80 6 9	.... 0 1 6
	£.2363 5 8¼	Av.£.0 3 6¼

Ewelm Hundred.

Name and Description of each Parish or Place.	Total Money raised by the Poor's-rates, and other Rate or Rates, within the Year ending Easter, 1803.			At what Rate in the Pound, for the Year ending Easter, 1803.		
Baldwin, Brightwell	£.491	8	0½	£.	—	
Bensington and Crowmarsh Battle	733	18	11	0	5	0
Berrick Salome	124	4	0	0	4	3
Britwell Prior	121	14	0	0	7	0
Chalgrove and Rofford	404	0	0	0	3	8
Cuxham	160	7	10½	0	5	6
Easington	6	17	2	0	0	10½
Ewelm	658	17	6½	0	8	0
Haseley, Great	191	14	7	0	2	10
Haseley, Little	121	12	11	0	2	6
Holcombe and Brockhampton	95	3	6	0	3	0
Nettlebed	326	8	0	0	8	3
Newington and Berrick Prior	350	1	1¼	0	5	0
Nuffield	287	17	1½	0	5	9
Swincomb	412	9	5	0	7	0
Warborough	615	19	9	0	7	11
Warpsgrove	12	10	5	0	0	8½
	£.5115	4	4¼	Av. £.0	4	10

Langtree Hundred.

Checkendon	£.481	14	5	£.0	6	0
Crowmarsh, Gifford	233	4	6	0	6	2½
Goring	974	8	9	0	7	0
Ipsden	782	5	4	0	8	6½
Mapledurham	814	9	4	0	5	7
Mongewell	249	12	6	0	4	3
Nuneham Morren	340	15	4	0	7	6½
North, Stoke	189	4	2	0	6	4
Whitchurch	671	4	3½	0	5	3
	£.4736	18	7½	Av. £.0	6	3½

Lewknor

Lewknor Hundred.

Name and Description of each Parish or Place.	Total Money raised by the Poor's-rates, and other Rate or Rates, within the Year ending Easter, 1803.			At what Rate in the Pound, for the Year ending Easter, 1803.		
Adwell	£.47	9	0	....£.0	1	10
Aston, Rowant	427	3	0	 0	6	3
Britwell, Salome	195	7	11½	 0	6	0
Chalford	103	4	0¾	 0	6	8
Chinnor	624	13	10½	 0	9	2
Crowell	187	5	1	 0	6	3
Emmington	80	16	3½	 0	2	4
Henton	376	11	0	 0	9	0
Kingston, Blount	376	0	11	 0	7	0
Lewknor	346	3	11¾	 0	7	0
Lewknor-up-hill	284	17	4	 0	7	0
Postcomb	222	15	11½	 0	7	0
Sydenham	383	11	8½	 0	5	10
Stokenchurch	574	5	8	 0	6	0
	£.4230	5	10	Av. £.0	6	2½

Pirton Hundred.

Name and Description of each Parish or Place.	Total Money raised			At what Rate in the Pound		
Greenfield	£.339	16	2	....£.0	1	0
Pirton	1164	0	0	 0	5	6
Pishill	61	15	10¾	 0	5	0
Shirbourne	463	6	3	 0	5	0
Standhill	33	6	6	 0	4	6
Stoke, Talmage	79	10	9	 0	1	9
Watlington	1019	13	2½	 0	9	0
Weston, South	69	0	5	 0	0	6
Wheatfield	63	10	1½	 9	1	6
	£.3295	19	3¾	Av. £.0	3	9

Ploughley

Ploughley Hundred.

Name and Description of each Parish or Place.	Total Money raised by the Poor's-rates, and other Rate or Rates, within the Year ending Easter, 1803.	At what Rate in the Pound, for the Year ending Easter, 1803.
Ardley	£.63 15 4	.... £.0 1 9
Bicester, King's-end	112 17 5	.... 0 1 6
Bicester, Market-end	1281 13 $10\frac{1}{2}$	.... 0 6 6
Bletchington	496 10 5	.... 0 6 0
Bucknell	171 17 0	.... 0 3 0
Charlton	144 11 7	.... 0 4 0
Chesterton	202 3 8	.... 0 2 0
Cotisford	77 17 7	.... 0 4 0
Fencott and Mercott	189 14 $0\frac{1}{2}$	.... —
Finmere	222 17 $4\frac{1}{2}$	.... 0 4 6
Fringford	190 5 $3\frac{1}{2}$	.... 0 4 7
Fritwell	227 8 $1\frac{3}{4}$	.... 0 6 0
Goddington	144 0 1	.... 0 2 8
Hampton Gay	23 0 3	.... 0 0 8
Hampton Poil	75 1 0	.... 0 1 3
Hardwicke	21 0 0	.... —
Heath	90 16 $8\frac{1}{2}$	.... 0 2 6
Heyford, Lower	162 16 5	.... 0 1 $10\frac{1}{2}$
Heyford, Upper	173 15 11	.... 0 4 9
Islip	424 16 4	.... 0 5 2
Kirtlington	476 9 10	.... —
Launton	376 4 0	.... 0 4 10
Lillingstone Lovell	190 5 $3\frac{1}{2}$	.... 0 3 6
Middleton Stoney	410 17 $5\frac{1}{2}$	.... 0 6 8
Mixbury	500 4 3	.... 0 8 0
Newton Purcell	42 5 1	.... 0 1 10
Noke	110 3 3	.... 0 2 $4\frac{1}{2}$
Oddington	86 15 $2\frac{1}{2}$	.... 0 1 $4\frac{1}{4}$
Shelswell	25 19 8	.... 0 0 10
Somerton	272 10 $7\frac{1}{4}$	.... 0 3 8
Souldern	425 5 0	.... 0 1 10
Stoke Lyne	260 16 6	.... 0 1 $9\frac{1}{2}$
Stratton Audley	310 1 8	.... 0 4 0
Tusmore	24 0 0	.... 0 1 $1\frac{3}{4}$
Wendlebury	161 0 8	.... 0 2 0
Weston-on-the-green	168 8 $11\frac{1}{2}$	.... 0 1 8
	£.8338 15 $10\frac{1}{2}$	Av. £.0 3 $3\frac{1}{4}$

Thame.

Thame.

Name and Description of each Parish or Place	Total Money raised by the Poor's-rates, and other Rate or Rates, within the Year ending Easter, 1803.			At what Rate in the Pound, for the Year ending Easter, 1803.		
Ascott	£.92	6	10	£.0	2	0
Attington	25	17	10	0	1	4
Milton, Great	359	1	$3\frac{1}{2}$	0	4	$1\frac{1}{2}$
Milton, Little	261	8	5		—	
Tetsworth	433	3	6	0	4	3
Thame	3762	2	$5\frac{1}{2}$	0	11	6
Waterstock	48	8	0	0	1	0
	£.4982	11	4	Av. £.0	4	$0\frac{1}{2}$

Wotton Hundred.

Name and Description of each Parish or Place	Total Money			At what Rate		
Aston, Middle	£.114	11	6	£.0	3	6
Aston, North	262	11	3	0	4	0
Aston, Steeple	194	6	4	0	4	$10\frac{1}{2}$
Barford, Great (F)	327	10	0	0	6	0
Barton, Middle and Steeple	340	12	$4\frac{1}{4}$	0	4	$1\frac{1}{4}$
Barton, Westcott	147	8	4	0	4	1
Begbrook	57	18	$2\frac{1}{2}$	0	3	3
Bladon	391	18	3	0	9	6
Cassington	309	3	0	0	2	$5\frac{1}{4}$
Clifton	423	18	0	0	10	0
Coggs	366	13	$7\frac{1}{2}$		—	
Combe, Long	230	9	4	0	3	0
Deddington	1040	2	$10\frac{1}{4}$	0	10	0
Dunstew	437	4	10	0	4	6

Name and Description of each Parish or Place.	Total Money raised by the Poor's-rates, and other Rate or Rates, within the Year ending Easter, 1803.			At what Rate in the Pound, for the Year ending Easter, 1803.		
Ensham	£.1214	11	8	£.0	5	0
Glympton	116	16	$5\frac{3}{4}$	0	3	0
Gosford	33	11	2	0	2	0
Handborough	694	7	3	0	5	6
Hampton	264	8	7	0	10	0
Hensington	120	0	0	0	3	0
Kiddington, Nether	173	4	0	0	4	0
Kidlington	384	9	$8\frac{1}{2}$	0	5	$3\frac{1}{2}$
Leigh, North	700	6	0	0	8	6
Leigh, South (F)	210	14	8	0	1	6
Newington, South	607	9	10	0	8	0
Rousham	151	5	$2\frac{1}{4}$	0	3	10
Sandford	494	7	5	0	5	0
Shipton-upon-Cherwell	111	6	5	0	2	7
Stanton Harcourt (F)	401	16	5	0	3	$0\frac{1}{4}$
Stonesfield	270	13	9	0	1	8
Tackley	308	5	5	0	3	0
Tew, Great	459	0	0	0	3	$1\frac{1}{2}$
Tew, Little	148	6	$5\frac{3}{4}$	0	2	$4\frac{1}{2}$
Thrup	55	12	$10\frac{1}{4}$	0	3	4
Water Eaton	85	2	11	0	1	10
Woolvercott	297	3	0	0	4	$6\frac{1}{4}$
Wotton	1014	5	8	0	2	0
Worton, Nether	85	5	$3\frac{1}{2}$	0	1	$6\frac{1}{2}$
Worton, Over	55	9	6	0	1	6
Yarnton	329	8	6	0	5	0
	£.13,431	16	$0\frac{1}{2}$	Av.£.0	4	$4\frac{1}{4}$

Oxford

Oxford City.

Name and Description of each Parish or Place.	Total Money raised by the Poor's-rates, and other Rate or Rates, within the Year ending Easter, 1803.	At what Rate in the Pound, for the Year ending Easter, 1803.
Oxford	£.4674 3 11	£.0 4 4

Liberties of Oxford City.

Binsey	£.119 8 11½	£.0 1 3½
Giles, St.	632 8 0	—
Johns, St.	43 13 4	—
Woodstock, New	335 9 0	0 0 6
	£.1130 19 3½	Av. £.0 0 10¼
Total of the county of Oxford	£.103,559 10 6	Av. £.0 4 8

Observations.

1. Returns from 284 parishes or places in the county of Oxford, have been received, and are entered in this abstract. Returns from 273 parishes or places are entered in the abstract of the returns of 1785: 277 returns in the abstract of the returns of 1776.

2. Thirty-eight parishes or places maintain all, or part of, their poor in workhouses. The number of persons so maintained, during the year ending Easter 1803, was 1131, and the expense incurred therein, amounted to 12,124*l*. 8*s*. 8¾*d*. being at the rate of 10*l*. 14*s*. 4¾*d*. for each person maintained in that manner. It appears from the abstract of the returns of 1776, that there were then twenty-five workhouses, " capable of accommodating" 943 persons.

3. The

3. The number of persons relieved out of workhouses was 20,394, besides 2800 who were not parishioners. The expense incurred in the relief of the poor not in workhouses, amounted to 76,565*l*. 1*s*. 8¼*d*. A large proportion of those " who were not parishioners," appear to have been vagrants; and therefore it is probable that the relief given to this class of poor, could not exceed 2*s*. each, amounting to 280*l*. This sum being deducted from the above 76,565*l*. 1*s*. 8¼*d*., leaves 76,285*l*. 1*s*. 8¼*d*. being at the rate of 3*l*. 14*s*. 9¼*d*. for each parishioner relieved " out of any workhouse."

4. The number of persons relieved in and out of workhouses, was 21,525, besides those " who were not parishioners." Excluding the expense supposed to be incurred in the relief of this class of poor, all other expenses relative to maintenance of the poor, amounted to 91,024*l*. 9*s*. 8¼*d*., being at the rate of 4*l*. 4*s*. 7*d*. for each parishioner relieved.

5. The resident population of the county of Oxford, in the year 1801, appears from the population abstract to have been 109,620; so that the number of parishioners relieved from the poor's-rate, appears to be twenty in a hundred of the resident population.

The number of persons belonging to friendly societies, appears to be five in a hundred of the resident population.

The amount of the " total money raised by rates," appears to average 18*s*. 10¼*d*. per head on the population.

The amount of the whole expenditure on account of the poor, appears to average 16*s*. 8*d*. per head on the population.

6. The expenditure in suits of law, removal of paupers, and expense of overseers and other officers, according

ing

ing to the present abstract, amounts to 2614*l.* 19*s.* 3¼*d.* The amount of such expenditure, according to the abstract of the returns of 1785, was then 1496*l.* 12*s.* 6*d.*

7. The expenditure in purchasing materials for employing the poor, according to the present abstract, amounts to 1397*l.* 2*s.* 6*d.* The amount of such expenditure, according to the abstract of the returns of 1785, was then 289*l.* 14*s.* 9*d.*

8. The poor of seven parishes or places in this county, are farmed or maintained under contract.

9. The poor of the united parishes in the city of Oxford are maintained and employed under the regulations of a special Act of Parliament.

10. Eleven friendly societies have been enrolled at the Quarter-sessions of this county, pursuant to the Acts of 33 Geo. III. c. 54, and 35 Geo. III. c. 3.

11. The clerk of the peace states, that he " believes it is usual in the county of Oxford, to make the poor's-rate upon the rack-rental."

In five parishes or places in this county, the rate in the pound is stated on the rack-rental. The amount of money raised by rates in those parishes or places, is 2483*l.* 19*s.* 9½*d.* The amount of the rack-rental, as computed therefrom, is 9469*l.* 5*s.* 2¼*d.* Consequently the average rate in the pound on this rental amounts to 5*s.* 4½*d.*

12. The area of the county of Oxford (according to the latest authorities) appears to be 742 square statute miles, equal to 474,880 statute acres; wherefore the number of inhabitants in each square mile (containing 640 acres) averages at 148 persons.

SECT. V.——LEASES.

Mr. Freeman, of Fawley-court, gives leases of twenty-one years, from a conviction that his estate will be more improved under long tenures than it can be under short ones. To one farmer with a lease of fourteen, who had managed his land well, he gave an additional seven years. If such encouragement has not the effect which ought to attend it, all Oxfordshire farmers should with one voice condemn the culprits.

Mr. Percey—Suppose land rented at 20s. an acre as tenant at will, what is that land worth with a 21 years' lease?—*It is worth* 40s. The discourse at Mr. Fane's table turned on this main foundation of good husbandry; and without specific proportions, as in the answer to this question, all agreed that the difference could scarcely be estimated too highly: for it was well observed, what does it signify that a man has an ample capital in his pocket, if he be afraid to lay it out. Is there one single branch of his business in which the no lease will not operate?

Mr. Sarney—Let all England be farmed without leases, and tithe every where taken in kind, and England from that hour starves.

Mr. Sotham is of opinion, that not one farmer in ten would give 5s. in the pound more rent for a 21 years' lease, than he does as tenant at will.

Mr. Turner, of Burford, thinks that upon stonebrash farms a lease of 21 years is worth one-fourth more rent than such as are usually given, viz. 25s. instead of 20s. per acre.

Mr. Coker is against leases, and never gives them:
his

his argument is, that they tell the farmer when he can begin systematically to exhaust the farm safely to himself: and secondly, that in this rich country, there are no great expenses in improvements, and therefore leases are not wanted: if draining be wanted, Mr. Coker bears half the expense.

Mr. Coker is right, as to the encouragement, by bearing a proportion of the expenses of permanent improvements. The landlord endeavours to establish a confidence between himself and tenant, that his continuance may be ensured by skilful and liberal management*.

The value of a 21 years' lease on comparison with tenant at will, 7s. in 20s. Twenty-seven shillings instead of 20s. in the opinion of Mr. Edmonds: in other words, that the most ignorant and backward farmer would give this.

I had a lamentable account of this business in much of the country around Chipping Norton. None are granted; or what are next to none. The longest is six years, prescribing the six crops; but more commonly nothing more than an agreement, voidable in many cases at six months' notice. This incredible infatuation of landlords appears wonderful, and will gradually work effects little thought of. The system certainly retains great power in the proprietor's hands, but they pay severely for it, and is without exception the most expensive folly they can be guilty of.

" Banking of the ant-hills, and hollow-draining, are both very beneficial practices, but too expensive for tenants who have no leases, or who have not entire confidence in the justice of their landlords. Leases, though

---

* For this, consult the districts of Cleveland, and other parts of the North Riding of York.—*Mr. L. MS. Annotations.*

objected

objected to by many gentlemen, are a guard against events that time may produce to a tenant's disadvantage, or even ruin. Leases, rightly framed, and adapted to the circumstances of the farm, are, I insist, advantageous to both landlord and tenant."—*J. Chamberlin.*

Observation.—This is a subject, upon which a person in the least degree enlightened, must take the pen with much concern. Those covenants, so necessary for the benefit of agriculture, have been for some time going into disuse. Many considerable proprietors grant no leases; and others will consent to no better tenure than that of a single round of crops, or at most of two; and there are but few of the latter description. In the Essex Report, I had occasion to make so many observations upon this subject, that the less is necessary to be said at present. It must not, however, be dismissed without a few remarks.

Wherever any great and permanent improvements are necessary, the subject is very easily indeed dispatched: for without a long lease such improvements will never be undertaken; and that landlord who in such a circumstance refuses to grant it, lays a perpetual prohibition upon those exertions which can alone effect the amelioration of his estate. This is so obvious a truth, that it ought to be apparent to common-sense. Draining, wherever the wetness proceeds from springs, can be effected only at a very heavy expense. The formation of watered-meadows is done at a still more heavy charge. Improvement of fences is a business of great expense. Building additional and necessary conveniences, wanted upon so many farms, can never be undertaken by a tenant at will. The conversion of poor wastes into cultivated land, demands a variety of exertions:

tions: and all other great improvements are wrought at such an expense, that no farmer who has prudence, will undertake them without an adequate term for his reimbursement. All this is sufficiently obvious, nor have I heard it contradicted by any Oxfordshire land-lord of any knowledge in rural economy. But the more usual plea is, that such great improvements are rarely wanted in the county, and that for the common routine of the practice of the county, neither lease nor any extraordinary degree of confidence is wanted. In reply to this, I do not contend that leases are equally necessary in all cases. They are necessary in propor-tion to the sums that would be expended upon the land, were they granted; and they are unnecessary in proportion to the ease of cultivation without any extra-ordinary exertions. But in either case, the farmer may rest contented with inferior modes of culture, who with a lease would aim at very material improvements, though not of the species above enumerated. Every one who has practised agriculture, must know the ama-zing difference between a farm upon which every arti-cle of common good management is exerted, and ano-ther upon which the main object is to do and expend as little as possible: under the never-failing apprehen-sion, that it is at any time possible for himself to be re-moved, and another fixed to enjoy the fruits of his exer-tions. And who can be insensible to the difference of en-tering upon a farm, upon which every thing that ought to be done, has been done, and upon another upon which not one shilling has been expended which could with any propriety be withheld. Let this comparison be fairly estimated, and those gentlemen who are wil-ling to assert, that a long lease is worth but a few shil-lings additional rent, will find that they have not duly considered this subject. That there are ignorant, and entirely unenlightened farmers, who consider a low rent as the only object worth a moment's attention, is readily granted. Men will every where be found, who will not apply a long lease to its right purpose. Such men resemble the dog in the manger: by means of their lease they keep others out, who would do something, while they themselves do nothing. And were long leases to be granted at once to all the tenants of an estate, without discrimination, the landlord might possibly be most materially injured, and yet the land not be in the least improved. There is a right and a wrong way of doing every thing; and a careless landlord, or an ignorant steward, might certainly contrive to secure all the evils that can attend long leases, and fail to reap any of the benefits. Estates which are in such hands, are not likely to be improved, whatever system is embraced.

Had I heard in conversation such observations as these, " I think long leases a great disadvantage to land-lords, and therefore do not grant them ; but if any active, spirited tenant of mine was to make it appear to me, that it would be for my advantage to give him a long lease, I certainly should listen to him, and grant it." There would be some reason in the remark ; but I have heard no such thing : a general condemnation of the system is all one can get. " To grant long leases, is to give away your estate : it is to bind yourself, and leave your tenant free : it is to give them a knowledge of the exact time when they can begin to depredate, without injury to themselves."

These and many other assertions, about as valuable, are what have often met my ear in Oxfordshire. To enter into a detailed refutation of such arguments is not necessary.

necessary. A 21 years' lease may be signed without proper covenants in the landlord's favour. Such a lease may be suffered to run out, without the tenant's knowing whether the landlord intends to grant a renewal. Covenants which exclude a system of exhaustion in the latter years, may be wanting. A variety of circumstances bear directly upon these points; but let it be asked where are great and permanent improvements to be found? What estates have been most improved at the expense of tenants? and I might ask, where have estates been most raised, in their rentals? The cases which I could produce, of the effects of long leases, under proper covenants, would go near to do away a thousand such arguments.

Thus much for landlords; but tenants ought very much to be reprehended, when it is possible to say of them, that they would value a long lease but at a few shillings per acre, such as a fourth, fifth, sixth, or seventh of the rent, as tenant at will. If such men are really to be found, they are not those to whom I would give long leases : since it is impossible that such men can have employed their minds upon the improvements of which their husbandry is capable. This is a point which I would particularly recommend to the attention of the Oxfordshire farmers. Much as the length of a lease depends upon the will of the individual landlords, it is not altogether independent of the feeling of tenants : and so long as they permit the habit of their minds to creep on in a low estimation of length of lease, and the immense variety of improved cultivation in a very great measure dependent on it, they are not prepared for the indulgence which a different range of ideas might prepare for their acceptance. And this may be one cause for the custom of granting leases, so much

much declining. There is no greater nuisance than a long lease in the hands of a slovenly tenant. The landlord suffers severely: the land is no better for the lease; and perhaps the tenant is as poor with it as he would be without it. I am ready to grant this; but it proves nothing against long leases, judiciously given.

An Oxfordshire gentleman remarks to me, that his leases are granted for four years, and from thence from four years to four years, so long as both parties please; determinable on one year's notice. " My only restrictive covenant is, that if the land be cultivated differently for the last four years, than before, the tenant shall pay double rent. The object is to ensure a four-years' shift, but without restriction of crops. I think it good management in certain cases, to sow two white crops together, and afterwards two green crops, rather than adhere uniformly to alternate crops."

Covenants recommended for a Stonebrash Farm.

That the outgoing tenant shall leave one-seventh a layer of sainfoin not more than seven years old.

That one-seventh shall be a seed lay of two years, ready to sow wheat on, and the incoming tenant to have liberty to enter for that purpose; such clover having been sown with barley that followed turnips fed off by sheep.

That one-seventh shall be under seeds of one year, sown with barley, that followed turnips; and those turnips on the wheat, without oats intervening.

That one-seventh shall be a barley stubble, the incoming tenant having had permission to sow what seeds he pleased at the time of sowing barley: such barley to have followed turnips, and those turnips the wheat, without oats intervening.

That

That one-seventh shall be turnips following wheat on a two years' lay, on which seeds were sown with barley after turnips.

That one-seventh shall be a wheat stubble after a lay of two years, which seeds were sown with barley after turnips.

That one-seventh shall consist either of turnips (additional) seeds of one or two years, or of sainfoin, whether old or young ; no two crops of white corn having come together during the last five years.

That in case of seeds failing during any time of the lease, nothing to be substituted but tares for soiling, or hay (not for seed) ; and where this is not done, the course to begin again with turnips.

That in case of turnips failing, the tenant may sow on the fallow what he pleases ; but wheat not to be followed by white corn ; nor barley or oats by wheat.

If pease or beans be at any time sown, it must be in the place of a white corn crop.

N. B. This is the present state of cultivation in Oxfordshire: when improvements have been established, it may be relaxed, and these crops admitted when seeds fail ; but I have known farmers purposely sow bad seeds, in order to get pea crops.

No dung or compost to be carried off the land.

No hay, straw, stubble, stover, or turnips, to be removed.

Tenant not to occupy any other land.

Tenant to find all carriage for repairs, and straw for thatch, if used, and beer for workmen.

Sainfoin, if pared and burnt, to be before turnips, and then the prescribed courses to follow.

Suppose the fields of a farm, on a new tenant entering, to be :

One

One sainfoin seven years old, ... No. 1.

One an oat stubble after six crops, ... 2.

One a wheat stubble after five crops, ... 3.

One a two years' lay, ... 4.

One a one year's lay, ... 5.

One a barley stubble with seeds, ... 6.

One a crop of turnips, eaten, and ready for barley, whether sown by outgoing or incoming tenant, ... 7.

This ought and must be the case, if the farm has been managed on the stonebrash course of, 1. turnips; 2. barley; 3. seeds; 4. seeds; 5. wheat; 6. oats. The principle of the intended regulation should, or may be, to admit this husbandry for two rounds (in a twenty-one years' lease) or twelve years. To strike off the oats in the next round of five years; and particularly to regulate the last four years.

Hence at the end of the 17th year the steward must be alert.

———◆———

SECT. VI.—EXPENSES.

Mr. Turner, of Burford, is of opinion, that for 300 acres of land in the stonebrash district, two thousands pounds are necessary.

Mr. Newton, of Crowmarsh, is clear that five thousand pounds are necessary to stock five hundred acres of land.

Mr. Hairbottle, of Henley, is well convinced that the husbandry of that farm in his vicinity must be deficient, if stocked with a less capital than 10*l.* per acre.

In the Dorchester district, two thousand pounds consi-

considered as necessary for a farm of three hundred acres.

Around Blenheim, generally about 6*l.* per acre.

It requires 10*l.* an acre to stock a Chiltern farm, to do it well; but many are taken with less.

To enter and stock a stonebrash farm at Hayford, &c. &c. of 300 acres, demands from 1500*l.* to 2000*l.*

Thirty-nine years ago (in 1769), all Mr. Secker's King's taxes came to 1*l.* 8*s.* 6*d.*; at present they amount to from 30*l.* to 40*l.* At that time labour was from 10*d.* to 1*s.* per *diem* in winter; now 1*s.* 6*d.* Mowing grass was then 1*s.* 4*d.*; now 2*s.* 6*d.* Reaping was 5*s.* 6*d.*; at present from 8*s.* to 10*s.* 6*d.* The pay of a carpenter and wheeler then 1*s.* 4*d.*; now 2*s.* 6*d.* A horse-shoe 4*d.*; now 6*d.*

Mrs. Latham, of Clifton, who has attended to farming above forty years, thus stated the rise of times:

	Forty Years ago.				*At present.*		
Labour, per week, ...	£.0	6	0	.... £.0	9	0
Blacksmith, horse-shoe, ...	0	0	4	 0	0	6
A mason, per *diem*, ...	0	1	8	 0	2	6
A carpenterr, per *diem*, ...	0	1	8	 0	2	6
Poor-rates, ...	trebled.					

Compa-

*Comparison of the Expenses of Farming in Oxford-
shire, in 1790 and 1803; returned to the Board of
Agriculture.*

	1790.				1803.		
Price of day-labour in winter per week .... }	£.0	6	6	....	£.0	8	6
Ditto in summer	0	9	0	....	0	13	0
Ditto in harvest	0	13	0	....	0	16	6
Head-man's wages	8	10	0	....	11	5	0
Second ditto	7	5	0	....	10	0	0
Reaping wheat per acre	0	7	3	....	0	9	9
Mowing barley	0	1	5	....	0	2	0
Thrashing wheat per quarter	0	3	2	....	0	4	9
———— barley ditto	0	1	4	....	0	1	9
Filling earth per yard	0	0	$1\frac{3}{4}$	....	0	0	$2\frac{1}{2}$
———— dung per load	0	0	3	....	0	0	4
Blacksmith wheel-tire per lb.	0	0	$3\frac{1}{2}$	....	0	0	$3\frac{3}{4}$
———— plough-irons ditto	0	0	6	....	0	0	7
———— chains ditto	0	0	$6\frac{1}{4}$	....	0	0	$8\frac{1}{2}$
———— shoeing ditto	0	0	$5\frac{1}{2}$	....	0	0	$6\frac{1}{2}$
Carpenter per *diem*	0	1	8	....	0	2	5
Mason	0	1	7	....	0	2	5
Thatcher	0	1	7	....	0	2	$1\frac{1}{2}$
Collar-maker	0	1	10	....	0	2	6
Rise in rent per cent. 20*l.*							
Tithe per acre	0	4	6	....	0	6	0
Parish rates in the pound	0	1	$7\frac{1}{2}$	....	0	4	$4\frac{1}{2}$
Expense of an acre of turnips	2	6	0	....	3	2	0
———— barley	1	2	0	....	1	8	0
———— wheat	0	14	0	....	0	18	0

Labour

	Rise per Cent.
Labour in winter	30
———— summer	44
———— harvest	26
Head-man's wages	32
Second man's wages	38
Reaping wheat	44
Mowing barley	41
Thrashing wheat	51
———— barley	31
Filling earth	42
———— dung	33
Average labour	37
Plough-irons:	
Chains	26
Shoeing	18
Carpenter	45
Mason	52
Thatcher	34
Collar-maker	36
Average artizans	35
Rent	20
Tithe	33
Rates	169
Expense of an acre of turnips	34
———————— barley	27
———————— wheat	28

Labour

	Rise per Cent.
Labour	37
Artizans	35
Rent	20
Tithe	33
Rates	169
Average	58

CHAP.

CHAP. V.

IMPLEMENTS.

————◆————

THRASHING-MILLS are spreading in Oxford-shire, and there is merit in this point; but in all other respects, I have not visited any county which possesses so few implements deserving the attention of the pub-lic. There may be horse-hoes in the county, but I did not see one; yet I saw some within sight of Henley-bridge, out of the county : there are some drills, which will gradually introduce other tools which ought to be connected with them. Scarifiers and scufflers, on which so much in modern tillage depends, are very rare indeed.

Upon this subject, however, William Lowndes, Esq. remarks to me, that " the fashionable scarifiers and scufflers of London have been tried and exploded. Where the staple of the land is thin, and the subsoil uncultivated, it must be by degrees only that it can be rendered fit for cultivation, and that by ploughing occasionally deeper, and fertilizing the soil in the interim. The operation of the scuffler is to move the two sorts of soil, before fertilization of the subsoil, and hence the lands so used become a crop of weeds. The scarifier can scarcely be used in a flinty soil, and hence only the disuse of them. In pease and bean crops, both drill and horse-hoe husbandry are well known; in which are in common use, an excellent drill constructed in the neighbourhood, the **Westmoreland harrow** and

horse-

horse-hoe, of which latter I have been in possession, the principle of which was brought from Shropshire fifteen years ago."

Plough.—The ploughs most generally used, are the two-wheel one, the beam resting on a pretty high fore-carriage; and the one-wheeled plough, the beam low: both have straight mould-boards, or but little varied. Like nearly all the ploughs in the kingdom, they make good work when well held. The neatest and truest ploughing I any where viewed, was on the farm of Mr. Davy, at Dorchester.

W. Lowndes, Esq. and Mr. Newton, of Crow-marsh, have used double ploughs, on the Leicester-shire construction, with four horses or oxen. Those of the former, drawn by oxen, plough one acre and a half in six hours. They perform their work well, and the country labourer expresses no dissatisfaction.

The Rev. Mr. Filmer uses the Staffordshire two-wheeled plough, which makes good work; and on my desiring the ploughman to let go his hold, he did it readily, and the plough continued in its work long enough to shew a true construction, so far as ease of draught is concerned: but four horses were drawing them, where two without a driver would be sufficient.

" Our ploughs, though heavy, are conceived to be well suited to the nature of the soil in this part of the county, which is very deep and stiff. It is the received opinion here, that deep ploughing is neces-sary, which cannot be effected with a light plough and two horses; and the Suffolk iron plough has not been found to answer where it has been tried. We can plough an acre and a quarter a day *well,* with four horses; and we do not find that the Norfolk ploughs,

where

where they have been tried in this neighbourhood with
three horses, can do much more than half as much :
still the experiment cannot be a fair one, unless made
with Norfolk men."—*Marquis of Blandford.*

" It may be depended on, that farmers will not keep
or use more horses than are really wanted to do their
work to advantage. I brought ploughs from Leicestershire, ploughed with two horses; nearly killed
them; laid the plough by, over the hen-roost, and
then went to work in the old way."—*J. Chamberlin,
of Cropedy.*

Crane.—Mr. Fane has a very useful and simple
crane and windlass, for drawing up sacks to his gra-
nary.

Carts.—Mr. Hairbottle, of Henley, uses only two-
horse carts for all the business of his farm : he brought
them from Northumberland. I expressed an high
opinion of one-horse carts, and he readily admitted
their superiority.

This gentleman uses the Northumberland drill for
turnips on ridge-work, and a three-shared horse-hoe
for working the intervals: I have so rarely met with
horse-hoeing in Oxfordshire, which is so much wanting
in their bean culture, that I hope the practice will gradually travel from Henley.

Roller.—Mr. Turner, of Burford, has invented a
cutting roller, composed of twelve wheels, two inches
and a half thick ; and between them a space of two
inches and a half. They are three feet diameter. He
loads it so as to be sufficient work for six oxen ; passes
it over wheat after it is sown, or after it is up, and if
dry,

dry, cross and cross. In spring he has used it also upon wheat; it leaves the surface rough in diamonds, which he finds useful. The price twenty guineas.

Skim-coulter.—Mr. Kimber, of Little Tew, uses this tool, and finds it for certain objects very useful: his skim is fixed to a fore-coulter, which he finds to do the work much better than if attached to the common coulter, in which way some have had it, and could not work well. On hollow land, however, he does not at all approve of it, as slicing the surface of such, he thinks, much worse than turning it over in the common manner, and letting the surface vegetables be laid into the diagonal position, with an edging of them sticking perhaps out of the seams. In this notion, I conceive he is in an error. The vegetable growth upon the surface is to be rotted in some way, for it is ploughed in in every mode of tillage, and the only question is, how to rot it in such manner as soonest to convert it into the food of plants, so as they may be able to avail themselves of that food. This surely is the means of lessening in the greatest degree possible, that additional hollowness which must be caused in a measure, plough it in how you will. The furrow of common ploughing four, five, or six inches, is so shallow, that the warmth of the sun, the moisture of rain, and the influence of atmospheric air act to the bottom of it, and will convert vegetable substances into gaseous matter sooner than if exposed to the immediate action of the sun and air, which in all dry weather preserve them, and must occasion a greater hollowness than in the other case, in which they are sooner consumed.

The great success that has attended the system counter to Mr. Kimber's, could not have taken place if
his

The Oxfordshire Skim Plough.

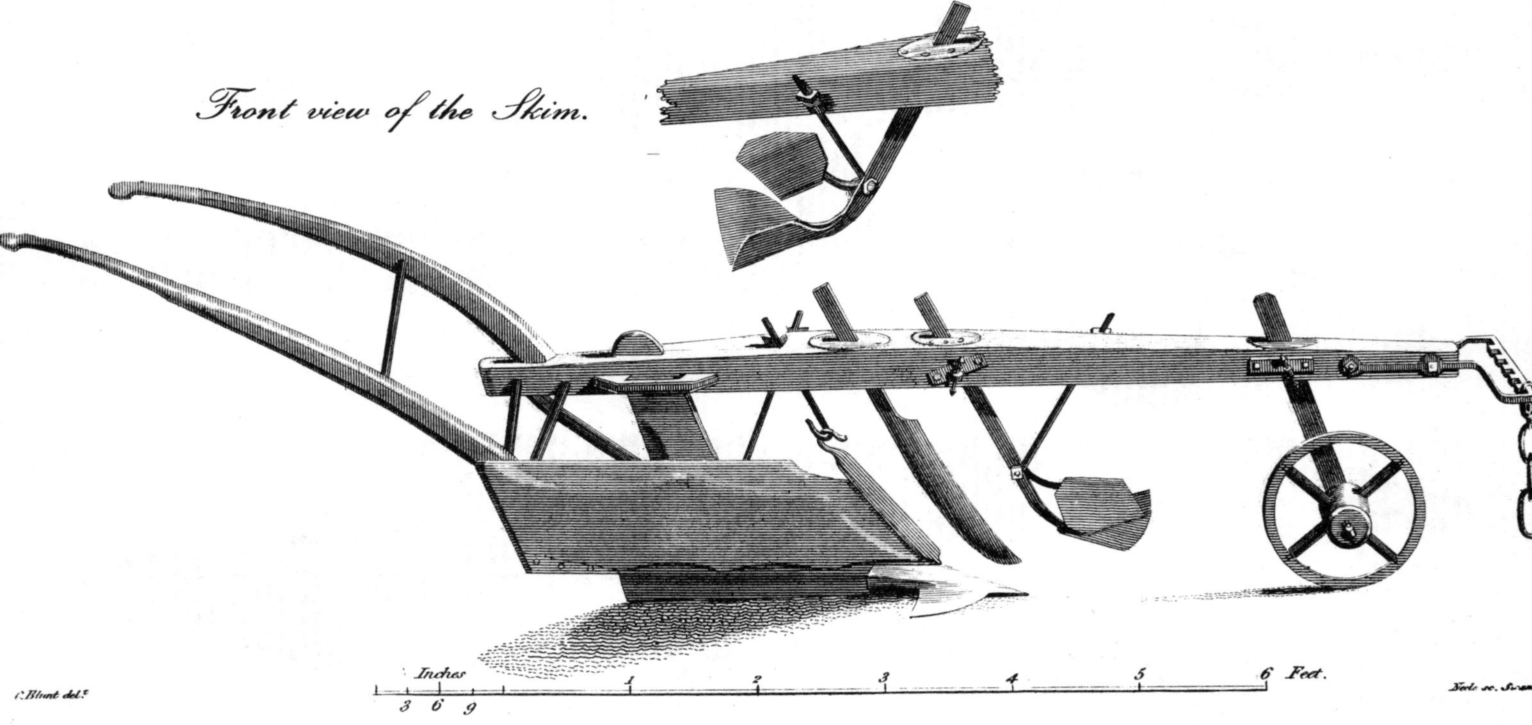

Oxfordshire Skim Plough. 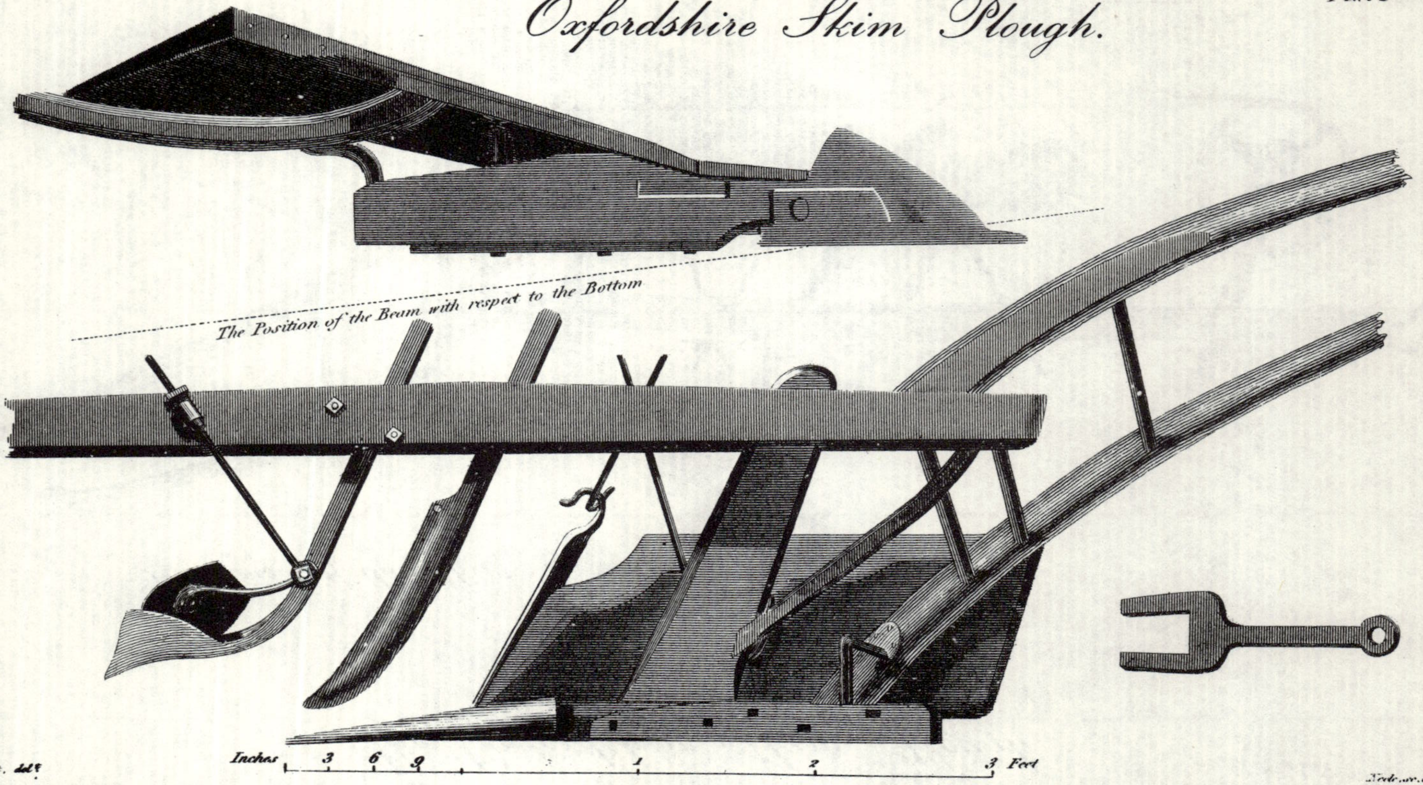
Pl. VI
The Position of the Beam with respect to the Bottom
Inches   3 6 9   1   2   3 Feet
C. Blunt. delt.
Neele sc. Strand.

Wiltshire Plough.
used in Oxfordshire.

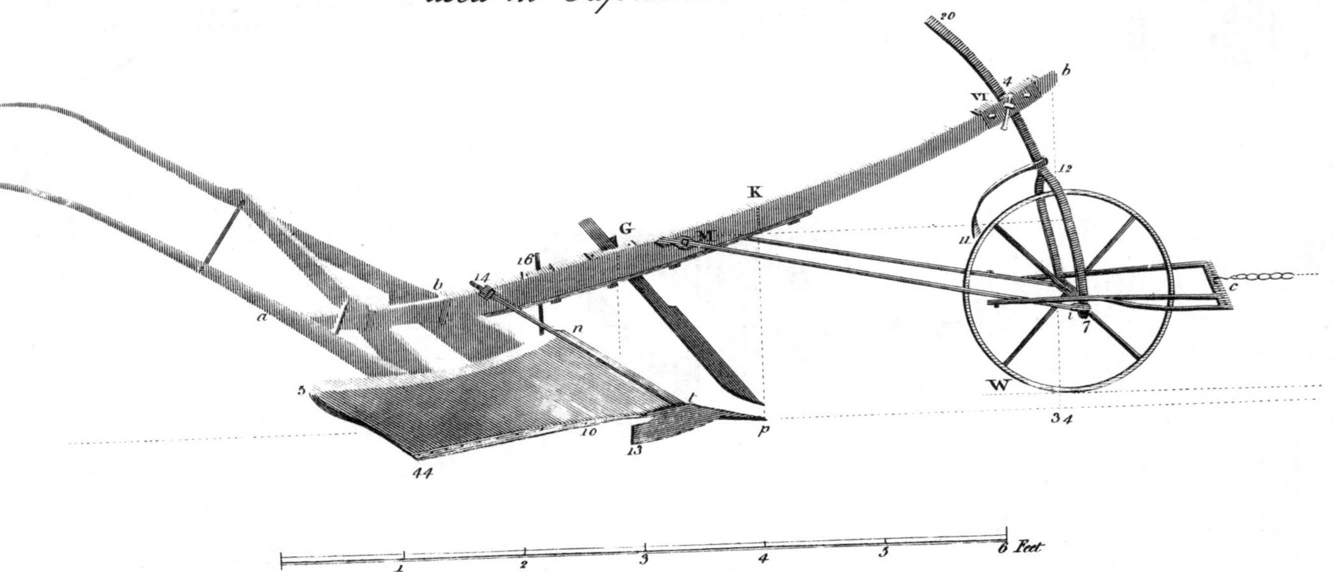

Oxford
Parts of Mr. Pullen's Wiltshire Plough.
Plate VIII 
Neele sc. Strand.
6 Feet
IX
W
M

his theory was just. Good farmers, when they skim-plough, or common plough hollow land, will adapt the operation to the soil the crop is to be put in, and the season of tillage; and by these means provide a remedy for the expected evil : but this is as clearly applicable to a skim as to no skim.

" The skim-coulter I have tried, but unsuccessfully, partly from mismanagement, and partly from the flinty nature of the soil. I use the Wiltshire eleven-share plough for various purposes : 1st, For cleaning the land, with tines constructed in various forms, fitted into the same frame; 2d, With shares for furrowing the land previous to sowing. The crops principally sown after this instrument, are vetches and turnips; but I intend to bring it into more general use, it being the best instrument, except for wheat and oats, that I have yet had in use. The reason why I do not think it proper to use it in those two cases, is, that for wheat, pulverization of the soil of my farm is ruinous. The closeness of its texture being requisite, the cultivation consists in one flat ploughing, and after sowing, a laborious sheep-treading. For the potatoe oat, which alone I grow, a similar process is necessary."— *Mr. L. MS. Annot.*

Two farmers at Great Tew, Mr. Busby and Mr. Neville, have lately got skim-ploughs from Mr. Kimber, of Little Tew, and they are much approved.

Mr. Kench, of Enstone, has an high opinion of the skim-plough, and especially for ploughing turnip land for barley, when the weather and treading have made it a little stiff, also if the fallow for turnips has become a little weedy; but does not use them on layers for wheat.—*Note.* He ploughs his lays deep for wheat.

Mr.

Mr. Wing, of Aston, was so obliging as to shew me his skim-plough at work on a clover lay for wheat; and I much admired the neat and perfect manner in which it did the work, going not more than three or four inches deep, and yet skimming off sufficient to bury every thing, and leaving seams, without any vegetation to be seen. Shallow ploughing clover land for wheat, they think essential on this stonebrash soil*.

Scuffler.—Mr. Wyat, of Hanwell, uses a scuffler which is good and effective: it has two rows of shares, five and four each, about six inches broad; the front ones cutting the intervals of the hinder. I saw its work on a bean-stubble of Mr. Salmon's, at Hardwick, and this work of a bean-stubble in Oxfordshire being new to me, after long travelling in the county, well merits an entry. It has two wheels, which rise and fall for letting it in or out of the ground. The land was preparing for wheat.

Thrashing-mill.—Lord Macclesfield has built one; it was erected by a mill-wright (Young) at Watlington; works with four horses, and cost, every thing included except the building which covers the wheel, 120*l.*

Mr. Juggens, at Wheatley, has erected a thrashing-

---

* It was with much pleasure I rode over the farm of Mr. Wing, at Aston. His skim-plough working a clover-lay very neatly for wheat; his stubbles of all sorts clean and bright; even that of pease and beans unusually clean; excellent crops of Swedes, and even of common turnips, for this generally failing year; very fine New Leicester sheep by twenty years breeding from the stock of his neighbour Mr. Creek; his seeds a beautiful plant; in a word, the farm abounded with every sign and proof of good husbandry.

mill,

mill, which cost him only fifty guineas ; a two-horse power. It thrashes every thing well except barley, and does that also by passing the straw through twice : barley, in raking together, has stones among it, which render it necessary to set the cylinder rather higher, and this carries the barley through so quickly, that two thrashings are not so expensive as many might imagine.

Lord Harcourt, at Nuneham, has a mill, a two-horse power, which cost 120*l*. It thrashes five quarters of wheat in a common day's work.

William Lowndes, Esq. has erected at Brightwell-grove a thrashing-mill, which was executed by Rastrick : it is worked by four horses : the drum-wheel, three feet and a half diameter, makes 260 revolutions in a minute, and, having 16 beaters, it gives 4160 strokes in that time ; a rake with four sets of teeth takes the straw, and delivers it to a second drum-beating cylinder two feet in diameter*. This is new to me, having never seen such an addition. Mr. Rastrick made the great drum with four beaters, but upon trial it was found that the straw was not clean thrashed, and four more added ; still it would not do its business clean, and therefore it was doubled, and now has sixteen. It was fortunately at work, and I observed the horses to labour considerably. They work four hours, in which time it does eight quarters of wheat. It thrashes every thing perfectly clean ; and does not break the straw more than

* This drum, termed the " dresser," turning in an opposite direction to the motion of the straw, beats it down, and in its descent strikes it against a circular board, faced with bars shod with iron, through the space of eighteen inches, by which the straw receives several additional strokes, and it is conceived has a great effect in dislodging that corn which has not been completely separated in passing the principal drum.

OXFORD.]   flails.

flails. Of barley it thrashes twelve quarters in four hours, and has done 16 in that time. The horses are not attached in the draught, which resembles pushing, by advancing with the lever before them, but in the common drawing way, the lever behind them, in which way Mr. Hodges, the bailiff, says they do much better. I numbered the sheaves as delivered, and found that it thrashed 43 in ten minutes.

Four horses a day's work, though only four hours, at 2s. 6d. per horse	£.0	10	0
Half a day of three men	0	2	6
Three women	0	1	0
One boy	0	0	3
	£.0	13	9
Suppose wear and tear	0	1	3
And interest of the first cost, suppose 200 acres of corn, at 4 qrs., 800 qrs.: if such a machine could be built for 300l. it is 4½d. per quarter, 8 qrs.	0	3	0
	£.0	18	0
Or per quarter	£.0	2	3

There can be no doubt at all of this answering. It dresses at the same time; and there is a chaff-cutter, and a corn-grinding mill with stones, for farm use, which are attached and worked or not, at pleasure. I have some doubt whether half-a-crown a horse be enough for the four hours' work; however, when it is considered that common thrashing is done at from 3s. to 5s. per quarter, and that here are all the advantages of clean work, and no pilfering, the benefit is plain.

The

The Bishop of Durham has built a thrashing-mill, which works with two horses, and thrashes five quarters of wheat in eight hours; nor is it harder work than strong ploughing for three horses. It cost fifty guineas, and five putting up. It thrashes wheat and oats very clean and well, but has not been tried for barley, which grain, I should fear, would require the fluted segment of a cast-iron cylinder to cover the top of the drum-wheel.

Mr. Kelsey, of Whitchurch, in 1806 erected a thrashing-mill, by Baker, of Exeter. The price forty guineas; carriage seven guineas: putting up with carpenters' assistance, and some beech timber, with a slight shed over the horse-wheel, made the whole expense about sixty guineas. It works with four horses, and by no means extraordinary hard work, if not fed too fast, or rather too thick.

Mr. Pearman, of Goring, has one by the same hand, which once thrashed twenty quarters of wheat in eight hours. But Mr. William Stone, of Inglefield, who has another of the same sort, has done no more than 14 qrs. of wheat in the same time.

Mr. Freeman, of Fawley-court, is erecting a most complete and very expensive mill, from which great things are expected ; two chaff-cutters are attached, one of which I saw at work: the quantity of hay mentioned as being once cut in a day into chaff by the two, is so great, that I shall refer to the persons concerned: I dare not repeat it.

Mr. Hairbottle has one, which was made in Northumberland, and brought to Henley.

The

The machine cost .... £.160	0	0
Carriage to London 10	0	0
Ditto to Henley 3	15	0
Men's wages putting up 7	10	0
Board, ten weeks 5	5	0
£.186	10	0

It requires four horses, two feeders, three women or lads to supply sheaves, two ditto to riddle, three men to remove straw, and one lad to drive.

It thrashes 30 qrs. of wheat in twelve hours, or twenty in a common day of eight hours. In twelve hours 45 qrs. of barley, or 50 qrs. of oats. If the corn be in sheaves it does its work perfectly clean; and Mr. Hairbottle binds all his corn into sheaves. The machines clear the corn from chaff, but it demands winnowing after. I saw it work with one feeder, and it thrashed 22 large and long sheaves in three minutes, without any alteration in the common movement of the horses.

The feeding-board is five feet four inches wide. The drum-wheel four feet four inches diameter, covered with sheet-iron, and has four beaters projecting four inches: it makes $88\frac{1}{2}$ revolutions to one of the horse-wheel; the horses going two miles and a half an hour in a path 27 feet diameter. The cogs of the wheels are of white-thorn, working in others of cast-iron; payed only with black lead; no grease. The level of the stage on which the men who feed, stand, is eight feet above the barn-floor.

Mr. Hairbottle thinks it utterly impossible to build a thrashing-machine that shall do justice to the owner, for any such sum as 50*l*. Success depends on firmness, strength,

strength, and solidity; not attainable for any such sum.

Mr. Green, at Rowney, near Henley; Mr. Pearman, at Yattenton; Mr. Good, at Kingston Sturt; and Mr. Jemmet, at Little Milton, all have thrashing-mills.

Mr. Knapp, at Kidlington, M. P. for Abingdon, has a thrashing-mill built by Freeman, at Stoney Middleton; it cost fifty guineas; a two-horse power: six quarters of wheat a day's work of the same horses.

Mr. Williams, at Woolworton, has another; it does not break the wheat straw so much as the flail; beans and wheat it thrashes well, but barley must be passed through twice. It does not dress; but heads seed clover extremely well.

Mr. Salmon, of Hardwick, has a thrashing mill built by Murray, of Durham. He saw one by the same artist at Mr. Lamb's, at Shelswell, near Bicester, which was so well approved of, that he employed the same maker: it was originally for two horses, and cost 50*l.* besides the expenses of carriage; but Mr. Salmon thinking the work too much for two horses, altered the horse-wheel to work with four, and it now does well and to his entire satisfaction. It thrashes every thing clean, with five beaters on the drum cylinder. I examined some wheat straw thrashed the day before I was there, and could not find a single kernel in the ears. The cog-wheels are all of cast-iron. It was built for the horses to draw behind in the common way, but now they are attached with the lever before them, as if in the attitude to push, which he finds to do better. It thrashes barley equally well with every other sort.

On an experiment, the horses going fairly, it thrashed five bushels of wheat in eighteen minutes, or sixteen quarters in eight hours; but the straw was very short,

and

and the wheat thrashed well; so that on an average of crops, no such quantity would be done.

Mr. Warrener, at Bloxham, has built one by the mechanic of Stoney Stratford: it works with a two-horse power, and thrashes six bushels of wheat per hour: and the same horses will do 48 bushels (nine gallons) in eight hours: it does its work clean; and cost 50*l*.

Mr. Miller, of Bampton, has one made by the mill-wright of Stoney Stratford: it is said not to work so clean as it ought to do.

CHAP. VI.

ENCLOSING.

THIS has been the capital improvement of the county; for proportionably to the extent of it, more land has been enclosed since I first travelled in it, which is about 40 years ago, I conceive, than in any county in England; but such has been the general inattention in preserving accounts of the various particulars which render such inquiries interesting, that most of the industry I exerted was in vain.

Waste Land Enclosed.

	Acres.
Fringford	230
Steeple Aston	188
Skipton	150
Black Bourton	1636
Westwick	750
Burford	90
Cropedy	800
Broadwell	800
Claydon	540
Tudmarton	500
Stratton	300
Ensham	500
Coggs	298
Adington	237
Brightwell	225

Melcomb

	Acres.
Melcomb	300
Stoke Lyne	1186
Burford	80
Westcot	200
Wigginton	1190
Hampton	204
Alvescot	600
Mollington	180
Ensham	1000
	12,559

There remain near 100 parishes unenclosed.

Enclosures in the first Forty Years of His present Majesty.

Enclosures	41
Acres (total)	68,480
Ditto, wherein no return of wheat	28,690
Acres of wheat before enclosing	4882
Ditto since	4770
Decreased	112

	Date.	Wheat Increase.	Wheat Decrease.
Fringford	1761	—	48
Merton	1763	—	100
Somerton	1765	50	—
Steeple Aston	1766	—	30
Westwell	1770	60	—
Cropedy	1774	—	100
Broadwell	1775	11	—
Claydon	1775	—	10
Ambrosden	1776	—	60

Alkerton

	Date.		Wheat Increase.		Wheat Decrease.
Alkerton	1776	....	40	....	—
Bointon	1777	....	—	....	50
Bucknel	1779	....	40	....	—
Stratton	1780	....	—	....	50
Little Tew	1793	....	10	....	—
Melcomb	1793	....	—	....	20
Westcot	1795	....	24	....	—
Hampton	1796	....	—	....	15
Mollington	1797	....	—	....	25
Kemscot	1798	....	50	....	—

Barley.

Enclosures	29
Increased in	14
Decreased in	13
As before in	2

Oats.

Enclosures	31
Increased in	29
Decreased in	2
As before in	0

Pulse.

Enclosures	25
Increased in	11
Decreased in	12
As before in	2

Cattle.

Enclosures	13
Increased in	9
Decreased in	1
As before in	3

Dairy.

Dairy.

Enclosures 18
Increased in 13
Decreased in 1
As before in 4

Sheep.

Enclosures 8
Increased in 8
Decreased in 0
As before in 0

Turnips.

Increased in 4

Potatoes.

Increased in 3

Artificial Grass.

Increased in 1*

If the standard quantity before enclosing be called 20, then at present it is,

Return of 17   enclosures .... 33 dairy.
Return of 5   enclosures 25

Acts passed for enclosure in Oxfordshire, in the first forty years of His present Majesty } 34

Acres 50,736
Acts wherein the acres were not specified 33
Acts in all 67
Returns made to the Committee of the House of Commons } 45

---

* Some of the returns of the Clergy named the object ; others did not.

Hampton

Hampton Poyle enclosed in 1796-7; saw a very great improvement in the tract that had been *lot* meadows.

The parish of Clifton, 39 years ago, was allotted by Mr. Hucks, being a private arrangement of his own. Each farm enclosed by an outline fence, but not subdivided. The seeds and turnips are fed off by means of hurdles, being all consumed by sheep. If stubbles are fed, a boy attends the stock.

Brightwell parish enclosed in 1800, under the same arrangement for fencing.

Mr. Latham considers hedges as very bad for corn, and had much rather be without them.

Burford was enclosed twelve years ago. It has not since produced so much corn; but infinitely more mutton and beef.

Fringford has been improved greatly in rent and produce since the enclosure, at least trebled in both. Stoke Lyne the same: no dairies in either. Mr. Bullock has enclosed Edgcot, 1100 acres, the rent of it will be trebled, and will all be pastured. Stratton Ardley was 500*l.* a year and now it is 2500*l.*; much of it will be under dairies. One estate there was offered for 3000*l.*, it is now 800*l.* per annum: great tracts have been laid down for cows.

Dodington, now enclosing, under the old rent of from 12*s.* to 15*s.*; will be 40*s.* all round.

Bicester 1400 acres, and 100 of common: in Wendlebury 1100, and 40 of old enclosure: no waste. The rent at Bicester was trebled; but at Wendlebury, being very good land, was hardly doubled; rise as from three to seven, it might be doubled. The Commissioners, Mr. Davis, of Lewknor, Mr. Davis, of Bloxham, and Mr. Collison, of Brackley. Land for tithe.—*Mr. Coker.*

Alvescot has been enclosed eight years, and very greatly

greatly improved, though the farmers are not the best to be found. A farm of 100*l.* a year while open, is now 300*l.* a year, and cheap. The vicarage was from 150*l.* to 200*l.*; now above 600*l.* a year. There were 600 acres of waste, most of which was pared and burnt. One farmer has taken corn incessantly, and found the bad effects. The land now recovered: the produce of the parish is certainly trebled.

Wotton has been enclosed about 37 years, and Mr. Sotham has not the least doubt of its having yielded full four times the produce in that time that it did in a like period before; and the rent is five times as much as it was in the open state.

Stonesfield about 500, of which 100 waste: soil stonebrash, with deep slate quarries.

Cassington enclosed about seven years, 2500 acres, of which 100 waste. The soil a cold clay, valued from 10*s.* to 12*s.* per acre: the home land a warm gravel, from 30*s.* to 35*s.* Meadows 30*s.* to 45*s.*

Ensham enclosed about seven years; there were from seven to 800 acres of waste, and about 4500 acres in all. Much of it a yellow wet loam qualitied at from 7*s.* to 10*s.* per acre; but the rest gravelly loam, from 25*s.* to 40*s.* Meadows from 30*s.* to 45*s.*

In the Heyford enclosure, 30 acres furze allotted for fuel; no cottagers had stock, only the proprietors; and therefore no allotments to the former.

In the Brightwell enclosure, a sum of money is paid by the proprietor to purchase fuel, brought by the occupier for the consumption of the cottagers entitled to common of turbary before the enclosure.

Expense.—Cassington cost, every thing included, 2*l.* 5*s.* per acre.

Stonesfield

Stonesfield about the same.

Ensham about 2*l*. 10*s*. per acre.

The fencing costs more than all the other expenses whatever.

Mr. Davis's bill, on all his enclosure, has not amounted to above 100*l*. per enclosure; though not attending so much as some, he may be less than others.

At a rough guess, from 4*l*. to 6*l*. per acre, including every thing.

I was informed by Mr. Taunton, that some enclosures have been made which have cost, every expense included, above 4*l*. per acre.

The total expense of Lower Heyford enclosure was 40*s*. per acre, for about 2000 acres.

Change of Management.—In Brightwell the principal change consists in the management. Before the enclosure the sheep were kept on the common from Lady-day, so long as food remained. Since the common has been converted into arable, and the farmers have substituted a breeding for a dry flock, in summer they are fed on vetches and rye-grass; in autumn on vetches and stubbles; and in winter on turnips. No other change has taken place.

Arable Produce.—It has been very greatly increased, in the opinion of Mr. Davis of Bloxham, who is clearly of opinion, that there is as much corn now grown on half the number of acres as there was upon the whole before; the other half turnips and grass, and the increase immense.

Enclosures all around Chipping Norton, have, in the opinion and information of Mr. Dawkins, unquestionably added greatly to the food of mankind; and the
country

country has been in consequence of them much improved.

Grass Produce.—Mr. Davis of Bloxham, has some
doubt if the produce of permanent grass-land be increased in consequence of enclosures keeping clear
of artificial grasses; for though many tracts have been
laid down, yet many also have been ploughed up.

Rent.—In general, rents have been increased by the
enclosures in Oxfordshire, reckoned at the first letting,
nearly double; and much more after ten or twelve
years.—*Mr. Davis.*

A farm at Hampton Poyle, belonging to Mr. Knapp,
which before enclosing lett at 175*l.* per annum, is now
400*l.*

Burford new enclosure was 3*s.* 6*d.*; now 25*s.*

Mr. Edmonds remarks, that enclosures have doubled
rents, and they are paid with more ease; and beyond
all doubt the produce of food for man is very greatly
increased by them; in some cases that of pasturage more
than corn; but it is very observable, that grass produce
is higher in price than corn.

They have now in some cases two-ninths given of
arable land in lieu of tithe.

The Farmer.—At Barton, the land was lett for
scarcely any thing, and the farmers generally as
poor as could be; enclosed, it lett at 20*s.* an acre,
and the farmers in easy circumstances, and doing well;
and in all of them the farmers in general very much
benefitted: nor have farms been more enlarged in consequence; they have been as much added to in open
as in enclosed. But it is a great error to suppose very
small

small farms beneficial; they are far from it, either to the community or to the occupier.

The Poor.—Very little difference to the poor; but not so much pilfering; far better for their morals; they never had the means of keeping cows, but when they had cottage common; the allotment much better, and the people in a better situation from the enclosure. —*Mr. Davis.*

SECT. II.—FENCES.

Mr. Davis is of opinion, that one very material way to lessen the expense of enclosures, is to have no posts and rails; which require no other difference of management than to lose stubble feeding for five years, which is a trifle, and to restrain the sheep from being without a shepherd. In Berkshire the farmers will not have them; they do much better with hurdles and shepherds. Mr. Davis did not post and rail his own farm, and never wanted them. Sheep will eat none, except while the shoots are young; on good land they are quite safe in five years.

Lord Macclesfield, on enclosing Shirborn, which he effected by only one commissioner, put out the fencing to Mr. Poulton, well known in this county and in Bucks, &c.; double post and rail fences, a ditch and bank, quick planting a double row, keeping clean, and supplying gaps for eight years, at 13*s.* the rood of eight yards; one-sixth of the money reserved; one-half of which sixth was paid in three years, and the rest at the end of the eight years. This is very reasonable.

I be-

I believe there is much truth in Sir C. Willoughby's idea, that open fields for corn are far superior to enclosures, supposing two circumstances: 1. That the lands be laid together; 2. That all extraneous rights be excluded ; and this turns on various points.

There being no other hedges than such as separate properties, is advantageous for corn upon all soils; for the freer ingress of sunshine and wind, the better: this benefit is in proportion to the wetness of the soil; and if ditches are wanted for drains, still these do not necessarily imply hedges. Farms without subdivisions, or at least without other than narrow grass communications, would force the occupier to practices that ought to be universal; such as mowing all grasses for soiling, and penning sheep as they are now penned in enclosures.

" That the greatest part of the improvement here spoken of, arises from enclosures, may perhaps still be very doubtful. In a dry country, arable lands lying in severalty will, *cæteris paribus*, be found to be more productive than when enclosed, where the trees without a doubt check the growth of the crops, and where the hedges harbour a quantity of small birds, which always do damage in corn, particularly if it be lodged, which it is more subject to be in enclosed fields than in large open fields, from the free current of air and wind being checked at particular points ; and though fences to the arable fields might be convenient, I do not know that any advantage would arise to the occupier, unless he should consider the hedge wood as an object, which never pays the expenses of setting, repairing, cutting, and making, &c. In the hilly part of this county, the parts which may perhaps be considered as waste lands, are in fact the very support of the farms, for they are

fine

fine healthy downs, which are occupied with the farmers'
sheep-walks, the bounds of which are as well known
by the shepherds and the common consent of the farmers, as if bounded by hedges : and it is the opinion
of most experienced and industrious farmers, that the
ploughing of such downs indiscriminately, would be,
after the first seven years, to the utter ruin of the
farms."—*Marquis of Blandford.*

" On a number of enclosures that have come within
my observation in the north of Oxfordshire, and the
country adjacent, I am well convinced that the products
of the earth are doubled by the event, and no diminution, but rather an increase of employment for the
poor."—*J. Chamberlin.*

Mr. Weyland, of Wood Eaton, having an allotment of 200 acres in the Islip enclosure, fenced it in
with a dry wall, five feet high ; the expense :

	s.	d.
Building 10*d.* per rod of five yards and a half, for one foot high	4	2
Quarrying five loads	5	0
Filling and carriage, including horses	3	4
In all	12	6

In the enclosures of Bicester and Wendlebury, Mr.
Coker's expense was in workmanship on the double
rows of posts and rails from felling the trees to railing,
2*s.* the pole of eight yards; under-banking 3*d.*; setting
quick 10*d.*, quick found. Ditch five feet 1*s.*; four feet
10*d.* to a spit wide at bottom. The whole expense of
completely making and preserving for seven years, by
contract, is 13*s.* the rod of eight yards. A ditch and
two rows of quick ; a mound on one side of the ditch,
OXFORD.]   with

with three posts and three rails; on the other a bank and post and two rails. If sheep are kept, the rails are drawn.

Stone walls at Wendlebury, &c. Quarrying the stone 6*d.* a square yard; building 1*s.* to 1*s.* 6*d.* a perch: they are very apt to shatter with frosts; but a foot thick in the middle laid with mortar, or even road dust made into mortar, holds them well together, and they are durable.

A General Enclosure.

" As it may be necessary to speak more minutely on this subject, I will begin first by stating the principal arguments which are used *against* enclosures; which are, that in proportion to the number of enclosures, the poor's-rates are increased, and the value of most articles of life enhanced; that a considerable quantity of land is taken up in making the fences; and further, that the expense of an Enclosure Act makes a great deduction from the probable improvement.

" In answer to the foregoing objections I would observe, that in regard to the poor's-rates, enclosures are not to be blamed; if properly regulated, they would have a direct contrary tendency, by enabling the farmer, under a better system of husbandry, to raise more corn, and keep more stock; the natural consequence of which would be, greater plenty and cheapness. And as there is in most parishes as large a quantity of downs, or waste land, to be broken up and converted to an arable state, as of corn land that is laid down to pasture after enclosing, of course, supposing the quantity equal, no deficiency of work arises to the labourers; but, on the contrary, more labour is required to cultivate the ploughed land on the improved system. **And**

And this conversion of down or waste into arable, may be allowed without inconvenience, on account of the grass-seeds to be introduced on the arable lands to maintain the stock.

" In regard to the land occupied by fences, supposing twenty-five acres in 200 thus taken up, such a quantity of land, when the fences are grown to maturity, will produce as much fuel as, before the enclosure, might arise from a worthless uncultivated waste of fifty acres, hitherto over-run with bushes and furze, which may now be ploughed up, and produce valuable crops of grain.

" But in the instance of stone walls, which are much in use in many parts of this country, the objection of a loss of land is trivial, as walls take up but a very narrow space.

" Or the objection of the damage arising from hedges may be entirely done away, by altering the clause in Acts of Parliament respecting the fencing, and leaving it optional to each proprietor either to fence his allotment, or leave it open after it is allotted to him *in several;* which was done in the last sessions of Parliament, in a Berkshire Enclosure Act, where there was a large tract of distant down land, on which it would have been difficult and expensive to raise a quick, and there were no stones at hand of a proper kind for walls.

" The expense of enclosing arises chiefly from the opposition of parties, or from bad management of the process of the business. Some parishes can be named, where the expense (exclusive of fences) has not amounted to more than one year's rent, and the advance of rent has been nearly triple immediately on the enclosure taking place. The improvement is found to be

greatest

greatest in convertible lands, of the stonebrash or
sandy kind, adapted to the culture of turnips and various grass-seeds; and in extensive downs, fit for tillage, and marshy wastes that improve by draining.

" Having spoken generally to the *objections*, it may
now be fair to enumerate some of the *advantages* arising
from enclosures.

" The first of these, is getting rid of the restrictions
of the former course of husbandry, and appropriating
each of the various sorts of land to that use to which it
is best adpated.

" 2. The prevention of the loss of time, both as
to labourers and cattle, in travelling to many dispersed pieces of land from one end of a parish to another; and also in fetching the horses from distant commons before they go to work.

" 3. There is a much better chance of escaping the
distempers to which cattle of all kinds are liable from
being mixed with those infected, particularly the scab
in sheep. This circumstance, in common fields, must
operate as a discouragement to the improvement of
stock; and it is a further disadvantage, that the occupier is limited both in regard to *number* and *kind* of
stock, instead of adopting such a number and kind as
are most suitable and proper.

" 4. The farmer has a better superintendance of
his labourers, when within the bounds of an enclosed
farm, than in an open field.

" 5. The great benefit which arises from draining
lands, which cannot so well, if at all, be done on single acres and half acres, and would effectually prevent
the rot amongst sheep, so very common in open field
land.

" 6. Lastly, the preventing of constant quarrels,
                                              which

which happen as well from the trespasses of cattle as by ploughing away from each other's land*.

" If by the increase of poor's-rates, be meant poor's-rates in general, it is a fact which cannot be denied : if those parishes only are meant which are enclosed, it is certainly wrong, as I have a knowledge of several parishes which are not enclosed, that have increased their poor's expenses in a much greater degree than any parish where an enclosure has taken place ; neither can the high price of provisions be attributed to enclosures, it having a direct contrary tendency, by not only bringing into cultivation large tracts of land which formerly lay waste, but by adopting an improved system for the cultivation of that which was before in arable, rendering it much more productive, both in feed for cattle and crops of grain, the natural consequence of which would be greater plenty and cheapness.

" Land of the stonebrash kind generally proves most advantageous on enclosing, which is to be attributed not so much to the nature of the land, as to the ignorance, bigotry, and self-interestedness, of the several occupiers, some of whom having never tasted the sweets of improvements, and knowing at the same time the present mode of cultivation will afford them a bare subsistence, are fearful of deviating from the old beaten track, and relinquishing a certainty for an uncertainty. Others there are of a sour, morose temper, who think all old ways best ; and if you speak of alterations and improvements, will say that their forefathers, who they think were as wise as those of the present generation, managed the field in the same way that it is at present, they therefore will not be deluded by promises of advantages, when perhaps it may prove

---

* Original Report.

their

their ruin. There are others also, of an envious, jea-
lous disposition, who will not agree to any alterations,
fearing that those who have more business than themselves will reap a proportionably great advantage by
the adoption of a new system. Now, those who are
unacquainted with the dispositions of common-field
farmers, would not suppose such absurd, inconsistent
objections could enter into the mind of man, more
particularly at a time that requires the utmost ingenuity
and industry of the occupier, to make a sufficient return, to defray the extra rent, and what is still more
felt, the enormous expenses of labour and maintenance
of the poor ; yet there is nothing more common where
there are a number of occupiers, than to have objections of the above kind made, when, perhaps, by adopting the turnip and grass-seed system upon their stonebrash land, and the strong land brought to three successive crops and then fallow, they might increase the value
of the produce nearly equal to an enclosure, at a time
when they are paying no more than a common-field
rent for their land."—*R. Wills*.

" All the advantages here enumerated are very great,
particularly the 1st and 5th. I have known years
wherein not a single sheep totally kept in the open field
has escaped the rot. Some years within my memory,
I am of opinion the rot has killed more sheep than
the butchers have. As one instance of prevention, I
occupy about 300 acres enclosed from the most rotten
part of an open field, where the rot often nearly cleared
the whole field. The first year was dry, and my sheep
took no harm ; I got the land drained and cleaned in a
short time, and have not lost one sheep of the rot since
I had the farm, which is now nineteen years."—*J.
Chamberlin*.

CHAP.

CHAP. VII.

ARABLE LAND.

———

SECT. I.—TILLAGE.

RIDGES.—At Wendlebury the ridges in the wet part of the old field are very high, and many are yet found in the enclosure. The Rev. Mr. Dupuis is very cautious of ploughing them down, as they contain only a cold clay ; but I found that others were gradually bringing them down, as he also has in some of his fields, but with caution.

In Aston and Lewknor fields, the soil rich strong loam, the ridges broad, and very little arched ; some nearly flat.

Sir Christopher Willoughby, thirty years ago, ploughed down some high ridges, in order to lay a field to grass, but has repented it ever since. The land has not yet recovered. He considers that the *staple* is the artificial child of cultivation ; and if it is buried, and the subsoil brought up by levelling, it is injured for an age. He did it for ornament, in forming a lawn ; but speaking as a farmer, he is decidedly of opinion, that old high ridges should on no account whatever be ploughed down.

He has often seen them laid to grass, in their old form, with great success, but scarcely ever when ploughed down. In such cases, if brought to good pasturage, it has been by force of manuring.

Plough-

Ploughing.—At Testworth they plough with four horses, and do an acre a day.

In the south-eastern division of the county, a party of farmers agreed that an acre is the general quantity that should be ploughed in a day by a team ; but many ploughs go out for three roods.   The expense thus reckoned :

	£	s.	d.
Four horses 1½ cwt. hay per week, at 4*s*., 6*s*. per horse,	£.1	4	0
One bushel of oats per horse, at 3*s*. 6*d*.	0	14	0
Chaff, 2*s*. per week,	0	8	0
Decline of value, 2*l*. 12*s*.	0	4	0
	£.2	10	0

	£	s.	d.
Per diem,	£.0	7	1½
Wear and tear,	0	2	6
Man and boy,	0	2	6
	£.0	12	1½

But on casting up the account, they observed that it was too low, for an acre of land could not be ploughed under 14*s*.

Horses—Are allowed at Wormsley, &c. 20 lb. of hay per *diem*, and a bushel of oats per week ; more for extra work. They plough with three or four, and do an acre a day.

At Adderbury either three or four horses in a plough. Some very few have made a trial of two.

Agreed by several very good farmers, that it is right to plough a stubble in autumn for a fallow, deep ; five inches considered as deep.

At and about Thumley they are of the same opinion, when they break up for the first time in May.

Mr.

Mr. Edmonds remarks, that beans do best on deep ploughing; but the after crops may suffer for want of due tillage being given to such additional depth. If deep, it should be continued.

If a man is going to leave his farm, he might plough deep for beans, and get a better crop by so doing; but the land for his successor might be injured.

At Great Tew, five to six horses per 100 acres arable: use four in a plough; sometimes only three.

About Wood Eaton, and its vicinity, from three to four horses in a plough, and this on soils not heavy. N. B. Mr. Weyland has an estate in Norfolk.

Mr. Cozins, of Golder, complains that his land ploughs extremely hard; four horses at length; eight inches deep for beans; and also in the first ploughing for a fallow. He keeps six horses per hundred arable acres.

Mr. Newton, of Crowmarsh, uses three or four horses in a plough; and for light work in summer, a double plough with four horses—the Staffordshire one. He ploughs deep for beans and turnips; but thinks that clover land cannot be ploughed too shallow for wheat.

The Bishop of Durham, at Mungwell, ploughs for most crops according to the staple of the ground, which is very shallow. He generally goes as deep as the staple will admit. His Lordship gained exceedingly by ploughing one inch deeper than the farmer before him had done; and which has gradually become an addition to the staple and the soil.

Some of the best ploughing I have seen in Oxfordshire was upon the farm of Mr. Pearman, at Maple Durham. The clover-lay was turned down for wheat, with furrows perfectly straight, and well turned. He gives three or four earths for turnips, and ploughs the turnip

turnip land twice for barley; for which grain, when sown upon a wheat stubble, three ploughings are given.

About Henley, three horses in stirring for turnips, but at other times four.

Mr. Creek, at North Aston, in breaking up a ray-grass layer, breast-ploughed it at 15s. per acre; left it for three weeks, then harrowed and broke it in pieces well, and ploughed: the ray-seed in the land grew, and was killed, and the wheat succeeded well. He spoke of this operation with much satisfaction, and answering better than the common way of merely ploughing, which does not get rid so well of the ray-grass.

We are apt to plough too much; the less we plough our land the better, said Mr. Percey: a good farmer will plough enough to keep it clean, and all beyond that is bad. The general feeling, that land may be ploughed too much, is to be noted in this remark.

Mr. Warrener, at Bloxham, has imported two ploughs (Rotherhams), and a ploughman to hold them, from Nottinghamshire; one for a pair, and the other for a single horse, instead of using the team of the country, which is three or four, and a driver. He ploughs an acre a day in common.

At Adderbury, on the rich sands, they do not plough more than four inches deep at any time, except on the little clay land they have, at the first breaking a fallow.

In the Dorchester district they use three horses in a plough, sometimes at length, and sometimes a-breast; but there are some very little farmers who, keeping only two horses, plough with them. They have three sorts of ploughs—a one-wheel plough, a two-wheel one, and a swing-plough. Mr. Thos. Latham thinks the one-wheel plough the best of these; and that it will

plough

plough the land in breaking up fallows, in seasons when the others would not work. His own expression was, a well-made one-wheel plough would break up the turnpike road.

Mr. Davy, of Dorchester, ploughs as deep as he can for beans; but as shallow as possible on clover-lays for wheat. He had a plough from Suffolk for two horses, but his men said that it drew harder than their own plough would have done, if used with the same team *.

Stale Furrow.—Mr. Tuckwell, of Cignet, is a great friend to putting in all corn crops on a stale furrow, especially on down or heathy land: it gives time for a consolidation of the furrow to take place for barley, as soon as possible after Christmas, and ploughs no more Whatever the crop, a stale furrow sure to be best.

Mr. Pinnal, of Westall, when I assured him that a pair of horses, or a pair of good oxen, would, in Norfolk or Suffolk, plough an acre a day, upon a soil full as strong as that I saw him ploughing for wheat, replied, that it might be so in those counties, but he was very sure it could not be done so here; and remarked, that Mr. Peacey, of North Leich, had a Norfolk plough, and gave it a fair trial; but found that it would not do, and threw the plough into a ditch. Without doubt, the report of such an experiment (whether the account be accurate or the contrary) will serve to prejudice these farmers in favour of their own

---

* We in general put in four horses, because it is the received opinion here, that deep ploughing is necessary, particularly the second and third tilth; for it is very common to plough shallow the first time to destroy the weeds, and we have not found that two horses will do upon strong land.—*MS. Annot.*

practice

practice, and confirm them in the idea, that four horses, worth 100*l.* are necessary for ploughing four inches deep, upon a loose friable stonebrash, the surface of which is as flat as any part of Suffolk.

Mr. Pinnal, of Westall, makes it a rule to plough very thin for wheat, nor does he like to plough deep for any thing; but deeper in winter-fallowing for turnips than for any other crop.

Breast-ploughing stubbles is a very common practice in this vicinity. Mr. Pinnal sowed 50 acres of turnips without any horse-ploughing at all, and without burning. It was a wheat-stubble, breast-ploughed in autumn, and the same work repeated for sowing the turnips: it costs 6*s.* per acre each time. The turnips proved a capital crop, as good as ever seen. He has also put in oats upon breast-ploughing only, and he cites this successful practice in proof of the propriety of shallow ploughing in general; for no horse-ploughing is so shallow as this work. He breast-ploughed and burnt a field, and then breast-ploughed again a part of it, horse-ploughing the rest, and there was no difference in the crop. A very material motive for shallow ploughing is, to have a firm bottom for the roots of corn.

About Thumley, on a stiff clay, or other soils very apt to *run together*, Mr. Reppington never knew any one attempt to save the spring-ploughings, and he thinks it would not answer.

Mr. Fane ploughed up a sainfoin layer in autumn, and left it untouched till the spring, when, without further ploughing, he harrowed in oats, and got ten quarters per acre.

But Mr. Sarney remarked, that he had found a fresh furrow, on a clover-lay for wheat, so superior to a part

of

of the same field ploughed five weeks before, that he much doubted if any general rule could be established. He preferred a fresh furrow for this grain.

Scarifying.—In preparing for turnips, Sir C. Willoughby saves a ploughing by the use of heavy drag-harrows, and finds it much better : ploughing will not answer the same purpose.

No scarifying, or scuffling, or shimming, or any operation, known about Baldon by the general run of farmers, except the common ones every where practised. They use very good drag-harrows, with four horses, and sometimes six on clay land.

There is not, in the vicinity of Baldon, the remotest idea of putting in any spring corn crops without spring-ploughings.

SECT. II.—FALLOWING.

Mr. Davis, of Bloxham, never saw any land upon which a naked fallow is necessary ; none even on the stiffest soils. He has been a practical farmer many years on a large scale, and has seen many counties minutely in his business, but was never yet convinced that any such necessity existed. Every one who knows this gentleman, knows that he has long been a Commissioner of many enclosures, having been employed upon 26 at the same time, and is an excellent practical farmer.

Mr. Cozins, of Golder, is a decided advocate for frequent fallows : upon the strong soils of his farm, he is of opinion that they must recur every third year, insomuch,

insomuch, that he fallows after beans, and clover, and vetches. To expose the land to the sun is of such importance with him, that it cannot be done away. In fallowing for wheat, he ploughs the land five times; not for freeing it from couch, but for the sake of letting the sun into it.

I have published cases, in which an extraordinary degree of fallowing has produced bad crops: Mr. Fane found a very different result; for having fallowed one year for wheat on his Chiltern-hill farm, and being disappointed in sowing it*, he continued the fallow another year, and then sowed wheat, and the produce was six quarters per acre.

———◆———

SECT. III.—COURSE OF CROPS.

Of all the circumstances in the management of arable land, in which the greatest revolution has taken place in the husbandry of this kingdom, none equals that which is to be found in the arrangement of the crops. Forty years ago, the knowledge of a true system was to be found only in the practice of Norfolk and Suffolk, and that solely on turnip soils; and the public having had little or no information of the fact, there prevailed throughout the kingdom such ignorance of what was right, and such a practice of what was wrong, that little could be expected till such a very material point was examined and explained. New ideas in this re-

---

* The field was sown with wheat; but partly from the severity of the winter, and from a large stock of hares and rabbits that lay very hard upon it, the wheat died. The field is a strong clay.

speot

spect have been spread, and agriculture has been improved in our counties pretty much, in proportion to the courses having been well or ill regulated. At present, we shall find in Oxfordshire something to commend, and something wherein improvements may be suggested.

Let us, in the first place, examine the courses which are actually found.

I. *Red-land District.*

In point of soil, this is the most interesting district of the county; but I am sorry to find, on reviewing my notes, that it is not so well applied as it might be.

In Deddington-field, now under an Act for Enclosure:

1. Fallow,
2. Wheat,
3. Barley,
4. Pease or beans, &c.

At Adderbury, the soil an exceeding fine, deep, reddish-brown sand on a gritstone; some of the very finest soil in the county. Mr. Wilson:

1. Turnips, manured with short dung, and eaten by sheep;
2. Barley on one earth;
3. Clover, commonly mown twice: Mr. Wilson feeds once;
4. Wheat;
5. Oats on two ploughings.

The addition of the oats is bad, and these coming in, no beans on such a soil is erroneous.

Mr. Bellow, at the same place, on rich sand:

1. Turnips,
2. Spring wheat,
3. Clover,
4. Wheat,
5. Oats.

While

While clover will succeed well once in five years, it is right to have it.   Why not beans between the wheat and oats ?

Mr. Davis, of Bloxham :

    1. Turnips ;

    2. Barley, ploughed twice, if the turnips be eaten off before Candlemas ;

    3. Seeds ;

    4. Ditto, once ploughed for

    5. Wheat ;

    6. Oats, two-thirds ; pease and beans one-third.

Whenever this most intelligent cultivator shall have time from his various employments, duly to consider his soil, he will without doubt make a variation in this course.

Mr. Warrener, at Bloxham :

    1. Turnips,   3. Clover,

    2. Barley, or spring wheat ;   4. Wheat ;

and there stops not, adding the oats commonly put in by his neighbours, as he does not approve of two white crops running.   This husbandry is certainly more correct ; but it does not draw from the land the greatest profit.   Beans on the clover, and then the wheat ; or beans after the wheat, and then oats, would either of them be an improvement.

Mr. Salmon, at Hardwick :

    1. Turnips,

    2. Barley, or spring wheat ;

    3. Clover, one year ;

    4. Wheat,

    5. Barley ; and sometimes adds,

    6. Oats.

When the last crop is taken, it only shews the vigour of the land.

                             Mr.

Mr. T. Payne, at Drayton :
1. Turnips, or cabbages;
2. Barley, or spring wheat;
3. Clover,
4. Wheat,
5. Beans, or pease;
6. Oats.

Much better than any of the preceding, and certainly has merit.

At Great Tew, on the fine estate of G. Stratton, Esq.; stonebrash; enclosed :
1. Turnips,
2. Barley,
3. Clover, mown once, and enough of it twice to supply seed for their own use;
4. Ditto, fed;
5. Wheat on one earth,
6. Oats.

When grass remains two years, to take beans, and follow them by wheat, would certainly be more profitable.

A great deal of the land in this district would without question support that highly profitable course of,

1. Beans, 2. Wheat.

If clover must be had, variations are admissible :

1. Beans, 4. Beans,
2. Wheat, 5. Wheat.
3. Clover,

And cabbages succeed so well, that the following would deserve attention :

1. Beans, 5. Wheat,
2. Wheat, 6. Cabbages,
3. Clover, 7. Oats.
4. Beans,

OXFORD.] *II. The*

II. *The Stonebrash District.*

The general characteristics of this district are, that the soil is dry enough for feeding turnips where they grow, and fertile enough for wheat; circumstances bearing immediately on the inquiry before us.

Enclosures, Fringford, and Stoke Lyne, &c. &c. :

1. Turnips,	4. Wheat,
2. Barley,	5. Oats.
3. Clover,	

Mr. Forster, at Bignal, stonebrash, a late enclosure :

1. Turnips,
2. Barley,
3. Clover; if at all foul, ploughed in the summer;
4. Wheat.

The Rev. Mr. Filmer's course :

1. Turnips,	4. Wheat,
2. Barley,	5. Beans, vetches, or pease ;
3. Clover,	6. Oats.

Also,

1. Turnips,	5. Wheat,
2. Barley,	6. Pease, beans, or oats.
3. } Clover, two years,	
4. }	

This in order to have more room for sheep. Another maxim is, not to have wheat oftener than once in six years.

Mr. Kimber, at Little Tew :

1. Turnips,
2. Barley, or oats ;
3. Clover, ray-grass, and white Dutch ;

4. Clover,

4. Clover, kept a second year;
5. Wheat, on one earth;
6. Oats.

At Sarsden :

1. Turnips,
2. Barley,
3. Ray-grass & clover,
4. Ray-grass and clover,
5. Wheat,
6. Oats.

They sow their wheat the beginning of harvest; and are now (in October) carrying out their dung from prepared heaps for turnips next year.

At Enston, and Westcot Barton, &c. :

1. Turnips fed by sheep,
2. Barley,
3. Clover for two years; first mown, second grazed ;
5. Wheat,
6. Oats.

Mr. Kench, of Enstone :

1. Turnips eaten by sheep,
2. Barley on one ploughing,
3. Seeds ; 12 lb. red clover, and two bushels of ray-grass; sometimes add trefoil, or white Dutch. Mow the first growth, feed the second ;
4. Ditto, fed ;
5. Wheat,
6. Oats.

At Salford, and Over Norton :

1. Turnips,
2. Barley,
3. Seeds, mown the first year, fed the second ;
4. Wheat,
5. Pease or vetches,
6. Oats. But some have their seeds only one year.

Another

Another course:

1. Turnips,   4. Wheat,
2. Barley,   5. Oats, &c.
3. Seed, one year;

But on the low land:

1. Fallow,
2. Wheat,
3. Beans,
4. Barley, or oats; and half the land sown with barley or oats, must be sown with seeds; and being once mown, is considered as a fallow, and ploughed for wheat.

Mr. Turner, of Burford:

1. Turnips,
2. Barley,
3. Seeds, mown once;
4. Seeds, fed;
5. Wheat,
6. Half oats, half vetches.

Mr. Tuckwell, at Cignet, near Burford:

1. Turnips,
2. Barley,
3. Seeds, ⎫ Very little mown at all, as they depend
4. Seeds, ⎭ chiefly for hay on sainfoin.
5. Plough once for wheat,
6. Oats, or pease, or vetches.

Mr. Secker, of Witney:

1. Turnips fed with sheep,
2. Barley,
3. Seeds mown, red and white clover, ray and trefoil.
4. Seeds, fed;

5. Wheat

5. Wheat on one earth.

6. Oats, or pease, or vetches; one-seventh of the whole under sainfoin.

Mr. Coburn, on clay:

<table>
<tr><td>1. Fallow,</td><td>4. Wheat,</td></tr>
<tr><td>2. Wheat,</td><td>5. Beans,</td></tr>
<tr><td>3. Beans,</td><td>6. Barley.</td></tr>
</table>

Duke of Marlborough:

1. Turnips, dunged;

2. Barley,

3. Clover one year, as the common grass is plentiful for feeding.

4. Wheat,

5. Oats.

Mr. Sotham, of Stonesfield, often sows pease instead of oats, throwing in turnip-seed immediately after ploughing the pea-stubble: this upon the land intended for turnips the year following. They yield some food for sheep late in the spring, and give seeds time to get a head.

The general fault of Oxfordshire husbandry, Mr. Sotham considers as the not having turnips and seeds enough, and sowing wheat upon barley-stubbles.

" But almost all over the whole district of the light lands of the north division, with which this county abounds, the following rotation is become very general in the enclosures:

1. Turnips,

2. Barley, with clover, rye-grass, or trefoil, or mixed;

3. 4. Clover, &c. as above, either one or two years, as the plant will continue;

5. Wheat,

5. Wheat, on once ploughing;

6. Oats, pease, or beans;

7. And in many places a part is continued in sainfoin till the plant is worn out.

" This course of crops, it is the general opinion, is
the best system that can be adopted for light or stonebrash soils, which cannot be fallowed too little; and
it has become a very general practice, on an enclosure
taking place, to divide each of the estates (after such
part is taken as is intended for permanent pasture or
meadow) into seven divisions, as nearly equal as may
be, for the purpose of adopting the six-round course
of husbandry before-mentioned, with the addition of
one for sainfoin, or grass-seeds; and then obliging the
tenant, by covenants in his lease, to follow the rotation
prescribed.

" In my opinion, there cannot be a better system of
cropping than is here stated, on soils of this description, if *the clover remains but one year*, and that
pease or beans (not oats) follow the wheat; except
changing the clover the third year for beans, and also
the pea or bean crop the sixth year for clover. Sainfoin also is highly proper."—*W. Dann.*

The reigning error in many of the preceding courses
is, the taking a crop of oats after the wheat. There
are soils of such fertility, as will yield barley or oats
after wheat without any material inconvenience (though
always better avoided); but the Oxfordshire stonebrash does not class with them. Another erroneous
practice is, letting the seeds remain two years, and
still sowing wheat. Ray-grass is very generally used,
which does little or nothing in preparing for wheat;
but admits pease or tares well, which, if duly cultivated,

vated, would prepare for the golden grain. As to the practice of sowing wheat on a barley-stubble, alluded to by Mr. Sotham, the farmers I conversed with do not commit so gross an error : it cannot be too much condemned. The best feature in the cropping of this district, is the quantity of sainfoin sown.

III. *The Chiltern District.*

This district has a near resemblance to that of Stone-brash, in the circumstance of doing well for both turnips, and wheat, and sainfoin.

On the Chiltern hills about Wormsley :

1. Turnips,
2. Barley,
3. Clover ; or trefoil and ray-grass ; and if to be kept three or four years, white clover added.
4. Wheat,

Those who are not so correct as others, add,

5. Oats, or barley ; pease, or vetches.

But if the seeds are left two or three years, then oats follow.

Mr. Dean, of English, on dry land :

1. Turnips,
2. Barley, or oats ;
3. Clover,
4. Wheat,
5. Turnips.
6. Barley, or oats ;
7. Trefoil, and white Dutch ;
8. Wheat.

In 400 acres of arable, he has only 80 acres of wheat. Therefore the above course is subject to variations. If he was to sow oats after wheat, he should get no turnips : as the oats would have eaten up the manure spread for the wheat.

At

At Maple Durham :

1. Turnips,	4. Wheat,
2. Barley,	5. Barley, pease, or oats.
3. Clover, or vetches ;	

The turnips are eaten by sheep, and the clover mown once or twice for hay.

Mr. Kelsey, of Whitchurch :

1. Turnips, or vetches fed off, and the turnips after them ;
2. Barley,
3. Clover, for one year ;
4. Wheat,
5. Barley,
6. Beans,
7 Wheat.

In defence of taking the barley after the wheat, he remarked that barley is *kind* after wheat, in this country.

By Mr. Percey, near Henley, on gravelly loam :

1. Turnips,	6. Turnips,
2. Barley,	7. Barley,
3. Clover,	8. Various articles at choice,
4. Wheat,	but not clover.
5. Oats,	

Also,

1. Tankard turnip,	4. Wheat, and the best
2. Wheat,	5. Pease, or oats, &c.
3. Clover,	

Mr. Freeman, of Fawley-court, has engaged a steward and bailiff from Northumberland (Mr. Foster), in order to introduce the system of agriculture practised

in

in that county.   He has also lett a farm contiguous to his own, to Mr. Hairbottle, from Northumberland. This gentleman brought a colony from his own county with him, both men and women.

The course Mr. Foster practises, and enforces among the tenantry by means of leases, is the following :

1. Turnips,
2. Barley,
3. Broad clover for one year,
4. Wheat, which is the favourite course of Northumberland.

This rotation is the well known Norfolk system ; and it is of such merit, where it can be regularly practised, that not a word can be said against it.   But neither Mr. Foster nor Mr. Hairbottle seem at all aware, that by clover being repeated every fourth year, the land grows what the farmers call in the East of England—*sick* of it.   This has been experienced in so many different districts, and under the most excellent management, that it is idle to form a doubt concerning it.

The agriculture of Northumberland may be called *new;* for I remember that county many years ago, in a state but emerging from the desolation of the *Borders*.   Clover is certainly a new article in it, and it may not yet have experienced the defect to which I allude; nor is it well ascertained, that the climate will subject it in the same degree to this misfortune as has been experienced in the South and East of England.   It is possible, that a greater degree of humidity may prevent the evil ; but this remains to be discovered.   In the vicinity of Henley, the case is abundantly different : the farmers there have the same complaint which has

occurred

occurred in so many other counties, and they have accordingly been forced to vary their crops with a view to this circumstance. These Northumberland gentlemen will not be many years before they feel the effect of it; and in that case, this regulated *system* must give way, or the effects will be extremely injurious.

Mr. Hairbottle has two courses: the first the Norfolk four-shift, noted above; the other is,

1 Summer fallow, or pulse;
2. Wheat,
3. Clover,
4. Oats.

Both manifesting his idea, that clover may be depended upon once in four years.

The most prevalent error in this district, as in that of the stonebrash soil, is the general practice manifesting no repugnance to having two crops of white corn in succession.

IV. *The District of Miscellaneous Loams.*

In the arrangement of the courses in this district, I shall throw them into two divisions:
1. On Bean Land.
2. On Turnip Land.

1. *On Bean Land.*—Bampton-field:

1. Fallow,   3. Beans,
2. Wheat,   4. Barley.

They practise, however, what in Oxfordshire is called *hitching* the field; that is, introducing variations by consent: and it is now proposed to substitute turnips upon half the bean land, then to have barley, and clover as a preparation for wheat.

Before

Before Hampton Poyle was enclosed, the open-field was under,

1. Fallow,   3. Beans,
2. Wheat,   4. Barley.

And now the change made is very uncommon, and is far short of what it ought to be:

1. Turnips, eaten on the land by sheep;
2. Barley,
3. Wheat,
4. Clover,   } which is a preposterous course.
5. Barley,

Another:

1. Fallow,
2. Wheat,
3. Beans, pease, or turnips;
4. Barley,
5. Clover, or vetches;
6. Wheat.

The fallow course on heavy land; the other for light good loam on gravel.

Garsington open-field:

1. Fallow,   3. Beans,
2. Wheat,   4. Barley, or oats.

In the arable lands of the grass district at Water-stock, &c. Mr. Ashurst, on clay:

1. Fallow,   3. Beans,
2. Wheat,   4. Oats.

On sandy land, of which there is some:

1. Turnips,   3. Clover,
2. Barley,   4. Wheat.

On

On the excellent deep loams between Stokenchurch and Tetsworth:

1. Fallow,
2. Wheat,
3. Beans.

In Tackley open-field:

1. Fallow,
2. Wheat,
3. Beans, or pease and clover mixed;
4. Barley.

Enclosed:

1. Turnips,
2. Barley,
3. Clover,
4. Wheat,
5. Pease and beans.

Mr. Cozins, at Goldar, open-field, but possesses enough to manage as he pleases:

1. Fallow,
2. Barley,
3. Beans.

1. Fallow,
2. Wheat,
3. Beans.

1. Fallow,
2. Wheat,
3. Clover.

1. Fallow,
2. Wheat,
3. Vetches.

In order to give a proper variation, the wheat course comes first, the clover one second, and the barley one the third; by which arrangement the clover is kept for spring feeding before breaking up for the barley-fallow.

If the shackles of an open-field forced these courses, nothing could be said against them but a call to enclose: as Mr. Cozins could vary them, I must consider the whole as a system of barbarism which ought to be exploded. Beans, clover, and vetches, to be in

every

every case followed by a fallow, is to adhere to maxims which the enlightened part of the farming world has reversed to their incalculable advantage. Such absurd courses are never found without assurances of their necessity : our soil is so stiff; our soil is so heavy ; our soil is so difficult, &c. &c. : the old changes are rung again and again, till new ideas expand into practice, difficulties vanish, and what were thought *necessities* dwindle into prejudices, that melt away before the rays, which animated efforts diffuse around. Mr. Cozins is active, spirited, and, in various other matters, extremely intelligent; he is in character well calculated for these efforts, and will, I doubt not, try them efficiently.

Upon the calcareous loam called chalk-loam, and also ragstone rubble, he has had,

1. Fallow and dung,   3. Beans,
2. Barley,   4. Wheat.

He is clearly of opinion that, 1. fallow, 2. barley, 3. clover, 4. beans, would not do, for want of more fallowing and exposition to the sun.

Mr. Buller, at Elsfield, enclosed strong land :

1. Fallow,
2. Wheat,
3. Beans,
4. Oats, or barley ;
5. Clover,
6. Ditto, fed in the spring till vetches; then
    plough for
7. Wheat.

But some of the clover ploughed up for winter vetches, ate off by sheep, and then fallowed. This is a very singular course, and merits attention.

On

On sand :

1. Turnips,   4. Wheat,
2. Barley,   5. Barley, or pease.
3. Clover,

Mr. Weyland, at Wood Eaton :

1. Turnips, dunged, and eaten by sheep ;
2. Barley, or oats ;
3. Clover,
4. Wheat,
5. Beans, or vetches, or pease.

In the open-fields here :

1. Fallow, and dung for   3. Beans,
2. Wheat,   4. Barley.

Mr. Rowland, of Water Eaton :

1. Turnips,   4. Wheat,
2. Barley,   5. Beans,
3. Clover, one year ;   6. Wheat.

Mr. Rippington, at Thumley, clay, enclosed :

1. Fallow,   3. Beans,
2. Wheat,   4. Wheat ;

but some sow spring corn on the fallow.

At Siddenham :

1. Fallow and dung,   4. Fallow without dung,
2. Barley,   5. Wheat,
3. Beans,   6. Beans.

This the old course at Crowell, but changed by Mr.
Kimber for,

1. Fallow,   3. Beans.
2. Wheat,

About one-third under clover by agreement, and
vetches.   His best wheat after clover.

On

On open-field, near Thame :

1. Fallow,   3. Beans;
2. Wheat,

on a very fine, reddish, loamy sand; and the crops great.

The Rev. Mr. Plaskett, at North Weston :

1. Beans,   4. Oats,
2. Wheat,   5. Clover, &c.
3. Turnips,

But intends to have his seeds lay five or six years, in order to be pared and burnt for turnips.

Wendlebury, enclosed seven years. Before enclosure :

1. Fallow,   3. Beans,
2. Wheat,   4. Barley;

and this course is still followed by some of the old farmers, who do not love change.

By the Rev. Mr. Dupuis :

1. Fallow and dung,
2. Wheat; average produce three quarters ;
3. Beans, four quarters ;
4. Barley, five quarters ;
5. Clover and dung, mown twice for hay ;
6. Vetches, fed, and then rolled down ;
7. Wheat.

At Bicester, on strong land :

1. Fallow,   3. Beans,
2. Wheat,   4. Barley.

Was so before enclosing, and some continue it. On the dry land now :

1. Turnips,   4. Wheat,
2. Barley,   5. Oats.
3. Clover,

The

The strong land much the most valuable, and, in the opinion of Mr. Coker, ought to be laid down to grass.

In Kidlington open-field :

1. Fallow : of late some vetches, ate off by sheep ;
2. Wheat,
3. Beans, pease, oats, clover, turnips, and a few potatoes and vetches ;
4. Barley ; part with clover, the rest fallow again.

And they have a large open field, upon which every man sows just what he pleases, which occasions such a confusion of headlands and abutments in tillage, &c. as can hardly be conceived.

Kidlington-field they do not fallow, as the mildew in that case is dangerous ; their course is very irregular :

1. Wheat,
2. Beans, or vetches ;
3. Barley,
4. Clover,
5. Wheat,
6. Beans, &c.

It is a remarkable circumstance, and shews how little beans are understood in Oxfordshire, that in all these courses, wheat following the crop is named but twice ; and the fact is, that beans in this county usually terminate the course, and are succeeded by a fallow—a sure proof that they are utterly ignorant in this branch of cultivation. Beans prepare for wheat (when well managed) to the full as well as turnips prepare for barley. But though they dibble many crops, it is across the lands, by short lengths of line. Horse-hoeing is excluded ; and though hand-hoeing is much talked of, I did not, even in this dry year, see a clean and well-ordered bean-stubble in the whole county, and only two horse-hoes. In every thing concerning beans, they want an entire revolution in their management.

2. *On*

2. *On Turnip Land.*—Mr. Lowndes, at Bright-well-grove :

1. Turnips,
2. Barley,
3. Clover, one half ; the other half vetches, trefoil or black-grass, pease, or turnips, as may suit ;
4. Wheat ; and as it is a new enclosure, this may be for some time continued without the clover failing.

Mr. Lowndes's general course over four-sixths of his farm is as follows : 1. turnips, 2. oats, or barley ; 3. clover generally over one half of the stubble, and over the other half vetches, pease, trefoil, or turnips : all the crops but pease fed off with sheep, and folded ; 4. wheat. As to the other two-sixths, ray-grass and Dutch clover sown after two years' green crops, viz. winter vetches, rye, and turnips, fed off with sheep in the fold, which introduces a conversion of management.

It is hoped that, as the farm consists of land in a state of waste till the enclosure, this course may, with occasional variations, be found permanently beneficial. Hitherto, the crops have bettered every year ; the clover is the least productive : the land lays too light and hollow, and loses stock in the winter.

Mr. Edmonds, on old arable loam on gravel, which was open-field :

1. Turnips,
2. Barley,
3. Clover honeysuckle, and trefoil, at Kemscott ;
4. Did two years, but now only one, as he intends,
5. Wheat ;
6. Commonly oats ; but Mr. Edmonds sows

OXFORD.]   vetches

vetches and pease, and ploughs immediately for stubble turnips, ate off by sheep.

Wheat has not done altogether so kindly (though the crops have been good) as if on one year seeds. Does not think the grain of so bright a colour as after one year seeds.

Mr. Edmonds occupies land, which, if wheat be sown on it, is raised off the soil by the frost, and fails : he therefore takes,

    1. Turnips,

    2. Barley,

    3. Seeds for two or three years, and pared and burnt for the turnips.

At Kemscott, on old arable loam on gravel, commonable ; loose and hollow gravel under the whole ; all turnip land :

1. Fallow,	3. Beans,
2. Wheat,	4. Barley.

Now, turnips, beans, seeds one year, wheat, oats.

Open fields at Baldons, &c. ; old course :

1. Fallow,	3. Barley, oats, &c.
2. Wheat,	

Also in Baldon-field :

1. Fallow,	3. Beans,
2. Wheat,	4. Barley, or oats.

Now :

1. Turnips,	4. Wheat,
2. Barley,	5. Beans.
3. Clover,	

Sir C. Willoughby's course :

    1. Turnips,

    2. Barley; produce, four to six quarters, and has had eight : this year six ;

                                3. Clover,

3. Clover, mown twice for hay;
4. Wheat, average produce; full three quarters more this year (1807);
5. Beans,
6. Barley, or oats; average produce, five quarters;
7. Clover,
8. Wheat, equal to the first;
9. Vetches, and, after being fed off, turnips;
10. Barley, or oats, and equal to the first.

A certain portion of the fallow in open fields taken off by consent for clover, tares, or turnips, tankards, to eat off early, and yet sow wheat: this is called a *hitch—hitching* a fallow.

In the improvements of Mr. Taunton, at Ensham:

1. Parc and burn for oats,
2. Turnips,
3. Oats,
4. Clover and ray-grass, mown.

He has had considerable success in the crops produced in this course. His farm consists of 250 acres, on which he keeps 200 Berkshire sheep.

Mr. Davy, a very capital and intelligent farmer at Dorchester:

1. Turnips,
2. Barley,
3. Beans,
4. Wheat,
5. Beans, or pease & turnips;
6. Barley,
7. Clover,
8. Wheat.

An excellent system, and, in Oxfordshire, does great credit to this enlightened husbandman. Pease are sometimes substituted for beans: there is great merit in this. An eighth is under a full crop of turnips; another eighth is turnips sown broad-cast amongst beans or

pease,

pease, hoed in as much perfection as hand-work can attain to ; Mr. Davy has had them two feet around, and usually very fair crops. Thus this important object on a sandy loam is secured. One fourth yields wheat, one fourth barley, one fourth beans or pease, and one eighth is in clover : more cannot be had, as the land is tired of it from repetition. The Norfolk system of, 1. turnips, 2. barley, 3. clover, 4. wheat, is a very good one, but not practicable where clover has been long cultivated : this is as near an approach to it as possible, and perhaps more profitable than the original system. Much must depend on the care and spirit with which the beans or pease of the fifth year are cultivated, respecting hoeing ; but so intelligent and clear-headed a man doubtless spares no exertions in that respect. Of course, in a large business, variations occur : thus, Mr. Davy sows trefoil, feeds it off till May, and then takes turnips ; after which barley, then beans. His stock of sheep proves that his stubble turnips must be good, for he has 600 sheep and lambs fed on a farm of 320 acres.

Mr. Thos. Latham, at Clifton :

1. Turnips ; all fed by sheep, except a very few drawn for cows ;
2. Barley,
3. Seeds, on about half the breadth ;
4. Wheat.

On the other half, beans or vetches instead of clover, by which means clover returns only once in eight years ; the clover land being in the next course beans, and the bean land clover. When the clover was sown every fourth year, it was very liable to failure ; but once in eight years stands well.

Mr.

Mr. James Welch, of Culham, upon red sand:

1. Turnips,
2. Barley,
3. Clover,
4. Wheat,
5. Turnips,
6. Barley,
7. Beans, or vetches;
8. Wheat.

Mr. Shrubb, of Bensington:

1. Turnips,
2. Barley,
3. Broad clover, one year;
4. Wheat,
5. Winter vetches, and turnips after;
6. Barley,
7. Beans,
8. Wheat,
9. Drilled pease, and turnips after them;
10. Oats, or barley;
11. Clover,
12. Wheat.

In this course, the clover stands well; and the turnips after the vetches and pease are as good as those of the first year of the course.

Mr. Bonner, of Bensington; open:

1. Fallow,
2. Wheat,
3. Barley,
4. Clover,
5. Wheat,
6. Beans.

Enclosed:

1. Turnips,
2. Barley,
3. Clover,
4. Wheat,
5. Barley,
6. Pease, &c.

Mr. Newton, of Crowmarsh:

1. Turnips,
2. Barley,
3. Clover,
4. Wheat.

Also,

1. Turnips,
2. Barley,
3. Beans,
4. Wheat,
5. Vetches, and turnips after.

He

He makes it a rule, by changing the rotation, to have turnips, clover, beans, and vetches; each of them return once in eight years. Upon very strong land he summer-fallows, and upon some of his light land will, after turnip-land barley, penn the stubble for oats. In this course:

1. Turnips,
2. Barley,
3. Oats,
4. Clover,
5. Wheat.

The Bishop of Durham, upon a calcareous loam, on a rough coarse chalkstone:

1. Turnips,
2. Barley,
3. Broad clover for one year; generally mown but once, but sometimes twice;
4. Wheat,
5. Tares, and some pease;
6. Oats, or barley; in which course clover will stand.

The turnips receive ten cart-loads per acre of yard-dung, each 50 cwt.; and the same dressing is given for wheat on the clover-lay. No summer-fallow.

But in some cases his Lordship takes,

1. Turnips,
2. Barley,
3. Clover,
4. Wheat,
5. Barley.

" The course of cropping in most parts of the county is so very irregular, that one may say there is no *regular* course in the county, where the farmers are not restricted. In some parts of the county, turnips are sown very forward, and fed off with sheep; and the land is ploughed, and sown with wheat by the middle of November, if the season suits. This practice is

esteemed

esteemed a good one; though it is universally allowed, that the wheat crop is in most seasons lighter, and more subject to blight, than with the common course. Fallowing is by degrees getting out of use, excepting for turnips, or upon very strong wet land, where they cannot be fed off; though upon such land, I think cabbages will be found to answer. I have this year about an acre, and there are many of them that weigh above 27 pounds each. There is not so great a check to good husbandry as suffering land to lie in common field; but land lying in severalty, in pieces of ten or twelve acres each, may, in many situations, answer as well as where enclosed. All wet land should be enclosed to drain it."—*Marquis of Blandford.*

SECT. IV.——DRILLING.

SCARCELY any drilling in the northern and eastern part of the southern division, but a little in the southern part of that district.

Mr. Fane has this year a crop of drilled turnips, done with M'Dougal's hand-drill, which for some purposes is a convenient tool.

Mr. Davy, of Dorchester, drills his pease with Cook's machine.

Mrs. Latham, of Clifton, drills every thing—wheat, pease, beans, tares for soiling, and even rye for sheep-food.

There is a little drilling in the vicinity of Bensington, but it is not much approved, and does not spread; and from all the observations Mr. Bonner has made, and from the information he has received, he is much inclined

inclined to think that the broad-cast husbandry beats it : the soil a gravelly loam.

" In general, the land in the neighbourhood will not bear much ploughing for wheat : the best wheat is on one ploughing. Experience, therefore, teaches the farmers, that the drill cultivation, requiring pulverization, must be hurtful ; few farmers have been found to adopt it, and I have not continued the practice. My observation teaches me, that the best crops proceed from the broad-cast ; about two bushels and a half per acre, thick enough to exclude the blast, but not to fall, except from heavy rains. For all feeding crops, and for barley, except on flinty soils, I am inclined to think drilling would be beneficial to the neat, careful and attentive farmer."—*Note by a Gentleman in the South of Oxfordshire.*

The cleanest pea-stubble I saw in Oxfordshire was on the farm in Blenheim-park, in the Duke's hands. Regular and straight drilling, and the freedom from weeds, meritorious.

Mr. Newton, of Crowmarsh, who is esteemed a very excellent farmer, is no advocate for drilling : he thinks the broad-cast husbandry much the better for wheat.

The Bishop of Durham, at Mungwell, drills all his wheat, at nine inches : weeds the crops carefully, but hoeing not found necessary. He does not drill his barley ; as his tenant did it once, but found more difficulty to rake it together free from stones, than with broad-cast crops, the rows of stubble impeding the teeth of the rakes.

Messrs. Foster and Hairbottle drill all their corn, the barley and wheat at seven or nine inches, according to the soil ; upon good at nine, and on inferior soils at seven. Mr. Hairbottle is not decided that the

crops

crops of barley are greater than the broad-cast. Beans,
from 18 to 24 inches : Mr. Foster thinks 24 the better
distance ; Mr. Hairbottle 20.

In the sphere of Mr. Pratt's knowledge, drilling, except for pease, has not been found to answer on stonebrash soils ; but on deep loams it may be good. Pease
are drilled, and very successfully.

The Rev. Mr. Filmer has drilled pease and beans
with the Kentish one-row drill six years ; and the
stubbles I saw were straight rows, and well done, and
earthed up : two shimmings and one earthing, if season,
and crop, and time admit : produce three quarters on
the average. Reaps and lays on the gavel, and binds
with tar-rope. The common way is, to mow, and bind
with the beans themselves.

Mr. Carr, of Beckley, and Mr. Butler, of Elsfield,
both drill ; and there is a scattering of this husbandry
from Islip to Wheatly. Mr. Butler, at Beckley New
Inn, is in this husbandry also.

Mr. Turner, of Burford, drills all his pease, and
hoes them twice.

Mr. Singleton, of Bampton, drills his wheat at nine
or ten inches, hoes it, and succeeds greatly. He much
approves this husbandry.

Mr. Coburn, of Witney, and Mr. Secker, drill all
their corn except what is dibbled : his wheat at nine
inches across the lands, as with beans ; some at twelve
inches, and yields heavier corn, but not so much in
quantity : he is clear his crops are quite as good as
broad-cast ones, with the advantage of the land being
kept clean by hoeing. Mr. Secker was the first who
drilled here ; uses the Vale of Evesham drill and scuffler.

Mr. John Wilson, of Adderbury, drills pease with
a small

a small drill attached to the fore-part of the plough, and moveable, so that the seed is delivered between the coulter and the horses, and consequently not trampled on: he does this with one plough, another following without a drill, consequently the crop comes up in rows at eighteen inches: keeps clean by hoeing, and wheat succeeds.

Mr. Davis, of Bloxham, drills every thing, and is much condemned for it by many neighbours: has drilled wheat for three years, and last year every thing. He uses Cook's and the Vale of Evesham drill: wheat, and other white corn, at seven inches; pease and beans, fourteen inches; and turnips, fourteen inches. Thinks, without doubt, that it is more advantageous than broad-cast; seed lessened two parts in seven. Success has been great. This year his wheat astonished his neighbours.

" With respect to the drill system, we cannot say that we think it will be found to answer in common, excepting for pease; and if any method be better for wheat than the broad-cast, it is perhaps the dibbling which is practised in some counties. And as it seems to be the object of the Board to employ the poor as much as possible, this method may be worth attention, which is done in a great measure by the children. Bean-setting is done by women and children, and we think preferable to the drill."—*Marquis of Blandford*.

The drill husbandry in this county must be considered as in its infancy; all that can be said of it in praise is, that there are some features which promise improvement.

SECT.

SECT. V.—WHEAT.

§ I. *Preparation.*—The preparatory year in Oxfordshire (Tillage and Manuring come into the following divisions) is occupied in clover or fallow; the courses of crops already detailed explain this sufficiently. Beans ought also to be very generally relied on for the same object; but the moment of that improvement is not yet arrived. Wheat, by bad farmers, is sown after barley, which ought never to be permitted.

In Brightwell, two farmers occupy bean lands, and both sow beans as preparatory to wheat; the rest use pease or vetches; no fallow.

§ II. *Soil.*—It is a mark of the general fertility of the county, that wheat succceds well on all the soils to be found in it.

The rich red land of Drayton, north of Banbury, is not reckoned by the Messrs. Paynes to be kindly for wheat; though not particularly subject to the mildew, yet the soil is too loose and hollow, and the crop liable to be root-fallen and damaged by frosts. I found Mr. Thomas Payne ploughing a pea-stubble pretty clean (deep for this grain), and expressed my wish to his brothers, that he would leave one stripe a few yards broad, which the ploughs had left unploughed, and that he would scuffle it well on the surface, and harrow in the seed. It did not want ploughing, and in such cases scuffling would for a loose soil be preferable. They appear to me all to be so free from prejudice, and ready to try experiments, that I think it very likely to be done. The soil yields fine turnips, barley, pease,

beans

beans and oats; but wheat not equal to the other products.

§ III. *Manuring.*—The almost universal management of the county is, to manure for wheat on strong
land; fallow and dung are relied on for the crop, even
where beans are cultivated; but turnips on dry soils
have a good share of the manure.

Mr. Newton, a very able manager at Swincombe,
dungs his layers intended for wheat in August; ploughs
and sows, and then runs the fold lightly over it. This
is found to do well.

This is not peculiar to Mr. Newton, who has had
that farm a very short time: it is more generally practised, and better done in Ewelm, where there are some
of the best farmers in the neighbourhood.

§ IV. *Tillage.*—Mr. Bonner, of Bensington, ploughs
his layers for wheat as shallow as possible; but a full
depth for turnips. If the land for wheat has been
dressed with rags, or penned with sheep, he is fearful
of burying the manure; but another reason for shallow
ploughing is, that the wheat loves a firm bottom to
root in.

Whatever the preparation may be (no fallows), Mr.
Davy, of Dorchester, never ploughs more than once
for wheat. He prefers sowing wheat on a stale furrow.

Mrs. Latham, of Clifton, remarked, that wheat has
often been tried on their gravelly loams on a fallow,
but never answered; nor are tares near so good a preparation for wheat as for barley, and for the same reason.

" A distinction must be taken between tares cut off
green

green by sheep, and tares for seed. In the former case, if sown so as to be cut in September, the wheat on one ploughing is productive. In the latter, the land is puffed, ploughs light, and is full of annuals, which impede the crop; and if the winter is various with frosts, the roots are exposed, and the stock is impoverished."—*Mr. Lowndes.*

At Bensington they plough their stubbles for a wheat fallow in May, give two stirrings, and the fourth is the seed earth.

" There are deep lands in this parish, to which only this observation alludes."—*Mr. Lowndes.*

Mr. Newton, of Crowmarsh, is of opinion that clover land cannot be ploughed too shallow for wheat.

Messrs. Foster and Hairbottle, in the Henley district, plough their clover lays five to six inches deep for wheat; the soil a good sandy loam on a gravelly bottom, and the crops not at all subject to be rootfallen.

Sir C. Willoughby ploughs his clover lays as shallow as possible for wheat, and especially if manured, whether by dung or fold, on sandy land, as the soil cannot be too tight.

Mr. Weyland, of Wood Eaton, ploughs a middling depth for wheat; about four inches on stonebrash.

Mr. Sarney, jun. of Sounds, on hill land, ploughed part of a clover lay for wheat early; he left it five weeks, and then ploughed the rest, and sowed the whole, and this was the best wheat throughout till harvest; and hence he clearly prefers it to sowing on a fresh furrow.

A party of excellent farmers, invited by Mr. Fane to meet the Writer of this Report, all agreed, that in ploughing a clover-lay for wheat, it should be as shallow

low as possible; and I found, in riding over the country, that this maxim seemed every where to be adhered to. If any reasoning be applied to it, it must
be that of leaving a part of former furrows solid for
the roots of the plants to fix in ; as all know that this
grain loves a firm bottom, and that too loose a one
causes a root-fallen crop.

Mr. Cozins, of Golder, thus fallows for wheat : the
first ploughing is given after the barley sowing is
finished, about old Lady-day ; this is done by *casting*
down the ridge : the second is begun as soon as the first
ploughing is finished, gathering up ; the third thwart
ing into large lands ; the fourth casting down in the
former direction ; the fifth gathered up, and the land
lays some weeks before sowing, the seed being harrowed
in a fortnight before Michaelmas.

In Bampton-field, they break the fallows for wheat
as soon as the barley is in, giving four earths in all ;
and ploughing in the seed on their broad ridges.

Mr. Kench, of Enstone, does not think it right to
put wheat in upon a shallow ploughing. He always
ploughs a lay deep, and even on his loose land ; and
his reason is, that slight frosts will lift it out of the
ground, which is prevented by ploughing deep, and
treading the surface. The climate on these stonebrash
hills is very cold, from their height. At Hook Norton
are two springs in the same ridge of country ; one flows
to the Thames, the other to the Severn.

Mr. Davis, of Bloxham, ploughs four inches for
wheat; but deep in autumn for turnips. He is now
(end of September) ploughing his oat-stubbles deep for
turnips.

Mr. Sotham, of Stonesfield, finds wheat to be best
on a stale furrow ; and thinks it cannot be ploughed
too

too dry, if possible to be done. He ploughs a full depth for wheat, as deep as four horses can draw.

§ V. *Putting in.*—Generally ploughed in on fallows, and harrowed in on clover : very little drilled.

Some of the farmers at Whitchurch plough their clover-lays for wheat, and scuffling in the seed, fold upon it; and this is thought very good management.

§ VI. *Season.*—At Bensington, on a gravelly soil, Mr. Bonner sows at New Michaelmas clover land first, that has been mown twice for hay: he is confident that clover mown twice, gives better wheat than feeding; two bushels and a quarter of seed on clover; two on fallow*.

The first rains that come in August, Mr. Turner ploughs, leaving it to sow the wheat on a stale furrow; but does not like to plough in dry weather for wheat.

Mr. Sotham, of Stonesfield, sows in August, if rain permits, and these are sure to be the best crops. Has had four quarters two bushels per acre, nine-gallon measure, on land his own property, over fourteen acres, sown the 26th August; manured with road-scrapings and dung. The hollow land is much subject to mildew, which also renders early sowing necessary.

Mr. Weyland, at Wood Eaton, on various loams, the end of September or the beginning of October.

The Rev. Mr. Filmer, on stonebrash, after the middle

---

* I heard the same thing admitted by a good farmer at White, in the Vale of Aylesbury, who was accustomed to feed his clover with ewes and lambs for the London butcher. The reason is, the increase of the tap-root of clover, so replete with the pabulum of wheat, when cut, which is not so when eat down.—*Mr. Lowndes.*

of

of August, the first rains that come, ploughs and sows; but if dry, ploughs it perhaps three or four weeks before sowing; then, when rain comes, harrows in the seed.

Mr. Kimber, of Little Tew, sows as early as the rains will let him—the last week in July, if there falls enough; and has had it green the 2d of August. The crops put in thus early, are sure to be, on an average of years, the best, and the freest from mildew in a year of that malady.

Mr. Kench, of Enstone, sows in or after harvest, on the first rain that comes. The cause of the last great mildew was the dry weather preventing early sowing, which is the best precaution against that distemper. He has some loose and hollow land, and on that pens the sheep as soon as it is sown, and drives them over the land to tread it.

Mr. Tuckwell, of Cignet, sows wheat as soon as the rains will permit him, even so early as the 26th of July. A stale furrow much the best, but hates to plough in dry weather for it. Should it be ploughed in too dry a state, he passes a heavy iron roller over it.

In Bampton-field, the soil a rich deep loam, about Michaelmas, which is full a month sooner than thirty years ago.

The times of sowing here minuted, are so entirely different from the practice in the eastern counties, that the most able and experienced farmers with whom I have conversed on the subject, are utterly at a loss to comprehend it. But as the same, or nearly the same contrast subsists in Oxfordshire itself, on the lower and deeper lands, it must result from an idea of the necessity of being so early, on account of the height and exposed situation of the stonebrash hills. We want

many

many experiments to be able to ascertain such a question; but who will make them?

§ VI. *Seed.*—About Baldon, two bushels to two bushels and a half, and three.

Mr. Newton, of Crowmarsh, sows two bushels and a half per acre.

Mr. Weyland, at Wood Eaton, two bushels to two and a half.

Mr. Cozins, of Golder, sows two bushels and one gallon before Michaelmas; two bushels and three gallons after.

The Rev. Mr. Filmer, at Heyford, has a very high opinion of changing seed. He has five quarters of Kentish red Lammas every year from that county, and as much Isle of Thanet barley, which supplies him with a sufficient change of both. His quantity per acre two bushels and a quarter, and as the season advances, two bushels and a half.

Mr. Tuckwell, near Burford, seven pecks in August (nine gallons three pints to the bushel).

Mr. Turner, of Burford, two bushels and a half on stonebrash: if sown late, three bushels.

Mr. Sotham, of Stonesfield, two bushels and a half to two bushels and three-quarters. There is no necessity for steeping, as old wheat is used. Last year Mr. Sotham sowed new wheat, not steeped, and the crop was smutty: that from the old seed had no smut.

On the rich sand at Atterbury, three bushels on clover-lays; but Mr. John Wilson has had five quarters an acre on fallow, upon stronger land, from one bushel and a half.

In Bampton-field, on their very fine loams, two bushels.

At Great Tew, two bushels and a half to three; three bushels and a half of barley; four to five bushels of oats.

Mr. Kimber, of Little Tew, sows two bushels and a half per acre (nine gallons), even so early as August; to allow for rooks, which at that season are apt to lay on it, and require a man with a gun to frighten them. He sows old seed, and, in that case, pickling is not necessary; though he has done it, as a greater security against the smut: but it is admitted that old seed is safe, especially if it be sown so early as August.

Mr. Cozins, of Golder, swims and skims the seed-wheat in brine; knew it once done in chamber-lye, and it destroyed the seed. If farmers would read books upon husbandry, they would know better than to deceive themselves by such practices: I published a case by Mr. Ellman, many years ago, which should teach them better.

§ VII. *Sort.*—White wheat with a red chaff, yields, with Mr. Sotham, the greatest crop, and sells at the best price.

Red-chaff Lammas the favourite at Wendlebury; but cone, though not an equal sample, is often more productive.

Mr. Lowndes, for the benefit of his neighbours, procures his seed-wheat from Taunton—the white Lammas, which being sown on his fresh land, is much improved, and well suited for the London market.

On the rich deep loams of Bampton, much white cone; also painted lady, which has white straw, red grain, and reddish chaff. They sow old wheat, and do not steep; but after Michaelmas, sow new. Cone wheat is not so liable to smut as other sorts; it is sub-

ject,

ject, however, to the mildew, but early sowing a great preventive : the distemper has been less common since the custom of sowing early ; and as hand-hoed wheat is more liable to it, all intended to be hoed should be the earlier sown. Minster much plagued with it.

Mr. Cozins, of Golder, in order that his crops may not be laid, sows Dantzic white, also red-chaff : he has tried likewise the white American ; but the straw breaks so easily, that it does not answer to litter his calves.

Mr. Turner, of Burford, has cultivated spring wheat, but not with much success : he does not like it, yet admits that it is not so subject to the mildew as common wheat.

Spring wheat at Stonesfield is sown at Lady-day. Mr. Sotham has no opinion of it, though certainly less subject to the mildew than the common sort.

Mr. T. Wyat, at Hanwell, sows it with great success on the rich red land of that district. He gets 30 bushels per acre, sowing three in April ; and the grass-seeds with it are better than with barley, which, after feeding turnips with sheep, is very apt to be hurt by the corn being too stout. He finds no objection to thus having wheat twice in a course.

Mr. Thos. Payne, at Drayton, in the same district, has cultivated this wheat with much success, and finds that the middle of March is a better time for sowing it than later. It yields as well as common wheat, that is, ten bags an acre (30 bushels), and it makes the very best of bread.

Mr. Bellow, at Atterbury, cultivates this grain upon the fine red sands of that parish with good success. He sows it the beginning of April, but did it one or two years later, and dry weather succeeding, the crop was

not

not so good. Another error he committed early in the cultivation was, that of sowing too little seed: he now sows three bushels per acre, and succeeds much better. He finds it to the full as profitable as barley, and, if the price of wheat be good, better, and all grass-seeds with it superior.

Mr. Warrener, at Bloxham, cultivates it very successfully. In 1806 he had 14 acres of it, put in on a clover-lay, sowing three, and even three bushels and a half (nine gallons) per acre, as he had found that less seed was with this grain unprofitable. His crops were very good, and yielding a greater value than barley, though he gets five quarters per acre.

Spring wheat has been cultivated by various persons in Oxfordshire for about eight years past. Mr. Buller, of Elsfield, has had good success with it on strong land; but on light soils it has been found apt to blight. His crops have been better than those of common wheat. He finds that this grain is the best of all others with which to sow clover.

Mr. Salmon, at Hardwick, had his best crop sown the beginning of March (three bushels); never found the frost to hurt it: sells as well as any other wheat: the seeds sown with it better than with barley, and never knew them to fail with it: has no objection at all to having wheat twice in a course instead of once, and gets quite as much spring wheat as of winter.

Jeremy Buck, at Ensham, had a small piece of Egyptian wheat, which produced in the proportion of ten quarters per acre.

Mr. James Payne, whose farm looks on to Lord Guildford's park, at Wroxton, has sown Cape wheat on the rich red-land of that district with much success: I saw it in the sheaf, and thrashed, and found it a very short

short round grain, very good wheat: he has had six quarters an acre of it, which no common wheat comes near to, for 30 bushels is reckoned a good common crop.

§ VIII. *Depth.*—In the Dorchester district, they do not plough above two or three inches deep for wheat.

Mr. Turner, of Burford, always ploughs shallow for wheat.

Cases in this article of depth might, from my notes, be greatly multiplied; but it may generally be observed, that the farmers in this county are friends to ploughing shallow for wheat, and depositing the seed at a shallow depth.

§ IX. *Drilling.*—Mr. Davy, of Dorchester, whose soil is sandy, drilled his wheat with Cook's machine for two years; but his crops were hurt by it, which he attributes to letting in the sun, and he left off the practice. There is very little drilling in the parts of the county Mr. Davy is acquainted with: less than there was.

Sir C. Willoughby drilled wheat with Cook's machine; but laid it by, from finding decidedly that it would not deliver seed enough. He has had it altered, and finds that it delivers the seed much better, though not yet quite sufficient.

For other particulars, see the article *Drilling* in general.

§ X. *Dibbling.*—The Bishop of Durham, at Mungwell, ordered an experiment to be made upon dibbling wheat. A very small portion of seed was used. His

farmer,

farmer, who is a seedsman of extraordinary accuracy of sowing, sowed broad cast, an acre by the side of it, using exactly the same quantity of seed. The crops were both great; something more than five quarters per acre, and the one exactly the same as the other.

§ XI. *Culture whilst Growing.*—Treading wheat well after sowing, Mr. Davy, of Dorchester, finds to be good management; he therefore folds sheep on all that he possibly can. His soil, sandy loam.

Sir C. Willoughby folds his wheat after it is sown, but not in wet weather.

Mr. Kimber, of Crowell, pens his sheep on the wheat after clover, as a preventive of the slug: they have no wire-worms.

Mr. Salmon broke up a grass-lay by paring and burning: the second crop was wheat; and finding the soil loose and puffy after he had sown it, he drove many cattle over it, till well trodden, and the crop proved one of the best he ever had.

Mr. Sotham, of Stonesfield, pens his wheat after sowing, which is necessary on hollow land.

In the Dorchester district, all the wheat, by good farmers, is hand-hoed in February or March, and weeded afterwards.

Mr. Sotham never knew wheat fed in Oxfordshire, though sown so remarkably early.

Sir C. Willoughby once ate off a crop with sheep, and it was lost: he will never again think of feeding wheat.

Mr. Davy, of Dorchester, has occasionally fed rank wheat, but not as a common practice. Hoeing he finds to improve thin wheat more than any thing.

Mr.

Mr. Sotham always rolls his wheat in the spring ; a practice found good on all land.

§ XII. *Harvesting.*—Nothing occurred in my inquiries relative to harvesting that is not common in every county ; but in the stacking of wheat, there is a spirit of exertion that merits great praise. I have seen round stacks more exactly built than some here ; but others are nearly faultless, and the oblong ones exceedingly well raised, perfectly well and neatly thatched, with straight and clipt edges ; very generally on substantial frames laid on capt stones, and the whole arranged at due distances in clean and neat stack-yards. They carry strangers into these yards, knowing well the merit of them. It is a gratifying sight, and especially to a Norfolk or Suffolk man, where the stacks resemble dunghills more than they resemble reeks in Oxfordshire.

§ XIII. *Distempers.*—Mr. Kimber is of opinion, that the mildew prevails more on the hills than in the vales.

Mr. Tuckwell, near Burford, has for many years found, that early-sown wheat is by much the most free from mildew and smut, and that the earliest sown is generally the best crop. He sowed a fifty-acred field with wheat with the same seed—part very early and part very late : there was no smut in the early sown, but some in that which was late.

At Tackley, the wheat being subject to the mildew, they sow in August : Lady Gardiner's bailiff gave this as the reason for early sowing.

In the open-field land of the two Miltons, the soil an uncom-

uncommonly fine gravelly loam, they are much subject to mildew, and on that account sow very early.

At Wood Eaton, they swim and skim in the common manner, and have no smut; but they are often troubled with the mildew.

§ XIV. *Produce.*—In Lewknor and Aston fields, seven to eight sacks per customary acre, which is to statute measure as eight to twelve, which is above five quarters.

The average of all the country south of Oxford, estimated at three quarters.

A field of the red sand at Dorchester produced six quarters and two bushels per acre, after the tithe taken.

The Bishop of Durham, at Mungwell, has $42\frac{1}{2}$ bushels per acre: his average produce about four quarters. When it is sown after tankard turnips, the produce is not equal to that after clover.

Average at Maple Durham, &c. three quarters and a half.

In the vale land at Whitchurch, three quarters and a half.

Mr. Weyland, at Wood Eaton, two quarters to two quarters and a half: soil, various loams.

With the Rev. Mr. Filmer, average produce of wheat 20 bushels.

The Commissioners' valuation at Kidlington, 20 bushels.

General average produce for some miles round Caversfield, on stonebrash, Mr. Bullock does not estimate at more than 20 bushels per acre.

At Enston, Westcot, Barton, &c. the average three quarters; they sometimes get five.

Produce at Great Tew, three quarters and a half.

On

On the rich red sand at Adderbury, Mr. John Wilson has had six quarters per acre, but never seven. Mr. Warrener, on the red land in the adjoining parish of Bloxham, gets four quarters on an average.

From all I saw and heard in the county, and combining the intelligence with the quality of the soil, I estimate the average produce of the whole county at three quarters; exceeding rather than falling short.

§ XV. *Price.*—The general idea in this county is, that if the average price of wheat falls short of 7*s.* 6*d.* per bushel, the farmers now must starve: and that they will not make a reasonable profit with that price, expenses have of late years risen so greatly.

§ XVI. *Straw.*—Last year (1806) straw sold at Oxford, at from 2*l.* 2*s.* to 4*l.* 4*s.* per load of 22½ cwt.

Four pounds a waggon-load about Stokenchurch.

§ XVII. *Stubbles.*—These are cut and gathered in the vales; but on the hills they cut closer, and therefore do not follow the practice; but I rode through many of their stubbles that would pay well for it, if taken early.

At Crowell and the vicinity, stubble sells at 2*s.* 6*d.* per acre.

Mr. Foster, at Bignall, whose stubbles I found very stout, gathers them for stacking in the farm-yard.

SECT. VI—BARLEY.

Course.—Mr. Goodchild, of Greenfield, is of opinion that the barley after wheat is better than after turnips. Wherever there are men who condemn cross-cropping, such opinions are to be suspected, if the soil is quite dry; but if the land be such as may easily be poached in feeding off, the fact often justifies the opinion.

Tillage.—The Bishop of Durham, at Mungwell, ploughs his turnip-land once for barley, but three times after wheat.

Mr. Singleton, &c. at Bampton, ploughs the bean-stubbles in October or November, and gives one earth more at sowing; turning in half the seed, and harrowing half in: three bushels to the field acre, or four to chain measure.

Mr. Weyland ploughs his turnip land twice for barley; sows four bushels per acre in April, and gets an average produce of from three to four quarters.

Mr. Secker, of Witney, has ploughed his turnip land once, and then scuffled in the seed, and thus got the best crops.

Mr. Latham, of Clifton, in common with all his neighbours, ploughs his turnip land thrice for barley. This tillage is thought necessary; nor do they seem to have made any trials sufficient to convince them that much of this tillage might be saved.

Mr. James Welch is in the same husbandry, but does sometimes put that crop in on two ploughings.

Mr. Pratt ploughs his turnip land once for barley,

turning

turning in half the seed ; and harrows half in upon the surface.

This grain is very rarely put in upon a summer-fallow in Oxfordshire, but it is by Mr. Cozins, at Golder, and upon a soil wet, strongly adhesive, and very difficult of culture. When I urged the propriety of sowing in the spring without spring-ploughing : *That cannot do : the soil under the frosted surface is so stiff and* CLUNG, *that it must be brought up to the air.* But that is of all others precisely the strongest reason for the husbandry. In proportion to the stiffness and harshness of the land that is under the surface, rendered friable by frosts, will be the difficulty of reducing it to a barley tilth : with rain it is mud ; with wind and sun it is clods as hard as iron, tumbled about by tillage over the dry moulds, turned down to be seen no more. The frost has given you a surface exactly prepared for the seed ; and by your own account, you bring up a new one precisely the contrary. If fallowing will not prepare the bottom of the furrow better than here described, I wonder much what the benefit of it can be which induces these gentlemen to consider it as so essentially necessary as to give it once in three years. He sows three bushels an acre ; ploughing in half, and harrowing the other half at top ; but not more than two turnings ; for the less harrowing these soils have, the safer.

Time.—At Stonesfield, a fortnight before, and a fortnight after Lady-day, is the principal time for sowing barley. Oats are sown three weeks earlier.

Seed.—Mr. Newton, of Crowmarsh, sows four to four bushels and a half per acre.

In

In the Dorchester district, from three bushels and a half to four bushels. The red sand has produced seven quarters and a half per acre, after tithe taken.

About Baldon, three bushels.

At Adderbury, on the rich red sand, three bushels and a half.

The Rev. Mr. Filmer, on stonebrash, three bushels.

Produce.—The Bishop of Durham, at Mungwell, in 1799, had six quarters six bushels per acre: his average about five quarters.

At Maple Durham, &c. five quarters; often six.

In the vale land at Whitchurch, five quarters.

At Enston, Westcot, Barton, &c. four quarters.

The Commissioners' valuation, at Kidlington, four quarters: the stonebrash barley there famous.

The average produce of all the country south of Oxford, estimated at five quarters by one intelligent husbandman, and at four quarters and a half by another.

At Great Tew, four quarters.

The general average around Caversfield, Mr. Bullock does not calculate to amount to more than three quarters and a half.

The Rev. Mr. Filmer, at Heyford, four quarters.

The nearest estimate to be given of the average produce of the county, is four quarters.

SECT. VII.—OATS.

Mr. Weyland, of Wood Eaton, generally ploughs twice for oats; but on stiff land only once, turning in

the

the seed half under furrow, and half harrowed in. About the end of March, five bushels of seed per acre. Crop, three to four quarters.

Mr. Davis, of Bloxham, ploughs but once, and in winter, so as to have the frost; and the same for beans.

Seed.—About Baldon, four bushels.

Mr. Newton, of Crowmarsh, sows five bushels to five bushels and a half per acre.

At Adderbury, on the rich red sand, four bushels.

The Rev. Mr. Filmer, on stonebrash, four bushels.

Mr. Fane has had, at Wormsley, ten quarters per acre, on the first ploughing of a sainfoin lay, on the rough, flinty Chiltern hills.

Mr. Davenport, of Shirborn, finds it more profitable to sow oats on strong land than barley, especially if it does not work so fine as he could wish.

The Bishop of Durham, at Mungwell, does not sow many oats, as his land is so well adapted to barley; but five quarters and a half per acre on an average.

Produce at Great Tew, five to six quarters.

Average around Caversfield, on stonebrash, four quarters.

At Enston, Westcot, and Barton, on stonebrash, five quarters.

The Rev. Mr. Filmer, on stonebrash, three quarters.

The Commissioners' valuation at Kidlington, four quarters; except on some deep soil: there, six or seven.

The general average of the county may be calculated at five quarters.

SECT. VIII.—PEASE.

On his lighter soils, Sir C. Willoughby sows pease instead of beans, but with attention not to repeat them too often; for of all plants, he finds that land soonest becomes tired of pease. There is likewise a practice of sowing what is called *poulse;* that is, beans and pease mixed, which produce good crops on the lighter lands.

Mr. Newton, of Crowmarsh, sows very few pease, thinking them a very bad crop for the land.

SECT. IX.—BEANS.

Sir C. Willoughby ploughs his strong bean land before the December frosts, and *sets* early in the spring; sometimes so early as January, dibbling on that one ploughing without further tillage: on light lands he ploughs in. This is important. The great discovery of late years in the Suffolk system, of banishing spring ploughings for spring crops, by such a note as this of management in Oxfordshire, receives confirmation; and it especially removes a common prejudice, which is so often met with, that such practices are and ought to be local, and that no conclusions can be drawn from one county for the conduct of another. He sets two bushels of seed per acre; always hoes twice, and some- times thrice.

Seed.—About Baldon, three bushels.

Mr. Davy, of Dorchester, dibbles his beans in part,

and

and drills the rest; hoes all well by hand, but horse-hoes none. By the way, horse-hoeing seems to be quite unknown in Oxfordshire. He prefers a stale furrow for beans, having often remarked the superiority.

Mr. Davy is one of the most intelligent farmers in the county, and has the reputation of being at least one of the best farmers in it; if the practice, therefore, of surface-working bean-stubbles was to be found any where, it would probably be on his farm, where are many beans; but the husbandry is quite unknown: one ploughing is given for the wheat.

One of the most interesting practices found in Oxfordshire relative to this crop, was the discovery I thus made in the Dorchester district, that in this county there are farmers who do not put the cart before the horse. I have travelled so long in open fields, and even in enclosures, with beans following wheat, instead of preceding and preparing for it, that it was a pleasure to find bean-stubbles ploughed up and lying in excellent order, ready to receive wheat-seed whenever rain came. For a stranger to recommend the practice of distant counties, is sometimes suspicious; but surely it must be entirely proper to refer these gentlemen to what is practised in their own county: let them come to Dorchester; they may take a lesson very well worth a long journey. Another circumstance, also, in this cultivation, equally merits the attention of such farmers; it is that of dunging for beans—a practice so many thousand times preferable to laying it on for the wheat. If wheat is to follow that crop, they always dung the land for the beans. The dung made from Michaelmas to Christmas is for beans; it is carried on in January; and being ploughed in, it is planted or drilled directly. The beans are hand-hoed twice, and

weeded

weeded once, if necessary. There seems to be but **one** deficiency in this culture, which is, the want of horse-hoeing : hand-work is but a trifling operation compared with that.

Mr. James Welch, of Culham, upon the red **sand** of this district, cultivates beans with much success; usually after barley, so as to enable him to avoid clover in a second round, by which means that grass **returns** but once in eight years. He has, with Mr. Latham, dunged for them, which being ploughed in, the beans are planted in February; and it is remarkable, **that** the cluster method of planting is preferred by him. Three, four, or five beans are set in a hole, nine **inches** asunder, and the rows eighteen inches apart : this **he** finds a much better method than planting single beans; at the same time, there is room for a good stroke of **the** hoe between the clusters, which is very effective in cleaning the crop. He hoes twice or thrice. The **poor** people pull the stubble, which is left quite clean, **and** is then ploughed for wheat. He prefers long dung **for** the crop, which lasts longer in the land than that **more** rotten.

They are all set at Bensington, at 5*s.* the **statute** acre; hoed twice, at 5*s.* each, and the same **for** reaping.

Mr. Shrubb, of Bensington, dungs his barley-stubbles for beans, when that crop succeeds.

Mr. Newton, of Crowmarsh, likewise dungs **for** beans. In common with all the farmers of Oxfordshire, he does not practise any method of cultivating the bean-stubble preparatory to wheat; nor did he ever **try** to put in wheat upon a bean-stubble without **any** ploughing. Pens the succeeding wheat after sowing. His quantity of bean-seed per acre, three bushels.

The

The Bishop of Durham, at Mungwell, has three pieces of land which do for beans; which crop he puts in after wheat, barley following the beans. He has them hoed with much attention, and sows turnip-seed immediately before the second of those operations, the crop of which is usually very good. The beans are set across the lands, in rows ten inches wide.

Mr. Cozins, of Golder, sets them across the ridges, and drills some by spraying by hand in the furrow; but in that case, is forced to plough so shallow, that it is a bad practice, especially as the ploughing afterwards is such stiff work; nor could he have wheat after beans, unless on his rubble land.

In the south district of Oxfordshire, nearly all the beans I saw were dibbled: 15 inches the common distance of the rows. They pay for it by the bushel, or by the acre: some 2*s.* a bushel, and others 4*s.* 6*d.* per acre. They are set across the ridges, to the utter exclusion of horse-hoeing; but all are hand-hoed. Mr. Kimber, of Crowell, and other good farmers, hoe all twice, and sometimes thrice; but the three hoeings only 7*s.* 6*d.* per acre. I did not see a clean bean-stubble in the district. *Query,* Is there one in the county? Most that I viewed were full of couch, and much other trumpery.

At Bicester, Wendlebury, &c. on all strong land, they set beans and hoe them twice: on drier soils, sow beans and pease broad-cast. Observing to Mr. Coker, that I had not seen a clean bean-stubble in the county, he thought it was for want of a third hoeing, which they will not give; but they have not a horse-hoeing idea, and never do any thing to the stubbles till they plough them. Never any dung for beans; all for wheat, if there be no turnips.

OXFORD.] The

The Commissioners' valuation at Kidlington, three quarters.

Some beans are drilled in the rich Thame district, but they are more commonly set.

Mr. Singleton, at Bampton, and other good farmers, plough the land only once, and that before Christmas; set the seed across the ridges, in rows 16 inches asunder. They are thought to put in too much seed (from three to five in a hole): Mr. Singleton thinks that one is better: always hoes twice, and he has done it thrice. They never spread any dung for them. Mr. Singleton approved the plan, and thinks it would be better to dung for them than for wheat, which is often found to be a bad practice, forcing the crop too much, and renders it apt to lodge.

Mr. John Wilson, at Adderbury, sows six bushels per acre, and has sown even seven; part of them pease, and never hoes.

The Rev. Mr. Filmer, on stonebrash, three bushels; and pease three bushels.

SECT. X.—TARES.

MOST of the soiling of horses that takes place about Baldon, is on tares, which they prefer to clover: some do it on clover also; but no cattle are ever soiled.

Sir C. Willoughby has had very good success with turnips after them; sheep are also fed on them.

Mr. Davy, of Dorchester, sows them for feeding sheep, and finds that they are a very good preparation for barley and oats.

Mr. Latham sows two bushels and a half per acre, a
week

week before and a week after Michaelmas : mows them for soiling horses in the stable; also for sheep; giving them in cage-racks, well contrived for the purpose, the sheep in a moving fold following close to the scythe; and finds this practice highly profitable. He finds tares an excellent preparation for barley; has had wheat after them, but not so good.

I was in no circumstance more surprised than to find, that among the very good farmers about Henley, they cultivate few tares. The Messrs. Perceys are of opinion, that this plant is apt to make the land too light and hollow, which, however good it may in many cases be for barley, causes the clover to die : for they find here, as in so many other districts, that this grass has been so often repeated that it fails. Here necessarily remains a question, whether the effect is to be attributed to the tares, or to such repetition? Autumnal rolling the young seeds deserves a trial.

Mr. Kimber, of Crowell, reckons tares very good for soiling; but thinks clover the worst of food for horses, though some soil with it. The first sowing of tares a month before Michaelmas, and then a month after; and spring tares in April, May-day, and the middle of the month : these yield much the greatest crops. He is now (in September) cutting the last spring-sown crop; but not good for horses so late.

Mr. Cozins, of Golder, feeds his horses in the stable with winter and spring tares; but the former yield most : he also mows them for his sheep.

Mr. Rippington, at Thumley, sows vetches for soiling his horses—a husbandry common in that vicinity : they are all spring ones, and sown in April, or the beginning of May, on a good manuring; and they get as good wheat after them as on land fallowed.

The

The Rev. Mr. Dupuis, at Wendlebury, has tried winter vetches, but they are apt to fail; and now confines himself to the spring sort, which he feeds with sheep.

Mr. Buller, of Elsfield, has an high opinion of vetches upon strong land, for they are excellent; but on sand, he thinks they *draw* too much; nor do turnips the same year succeed well after them. He soils horses on them, and *ties* them also. This husbandry, which is practised in parts of Oxfordshire, consists of staking the horses down with a chain of a certain length, so as to eat the crop with as little trampling as may be. I have seen it in Worcestershire, and must think it a most questionable practice, especially on strong land.

Mr. Sotham, of Stonesfield, finds that hollow land is bad for winter tares, and therefore ploughs the seed in.

Mr. Weyland, of Wood Eaton, has a high opinion of tares for soiling horses, but sows only the spring sort. Some tether the teams, which the Rev. Mr. Filmer does not approve of; and seeds such as are not wanted for soiling.

Mr. Turner, of Burford, considering that sheep, on a stonebrash farm, are the grand object, prefers sowing vetches after wheat instead of oats.

Mr. Bellow, upon his strong land, sows spring vetches for soiling his horses, on a part of his fallow for wheat; but if the land be not favoured in manuring, the crop is hardly so good.

Mr. Davis, of Bloxham, remarks, that tares are admirable for sheep, mown, and fed in racks—back into racks behind the hurdles; and after the tares, fallow in the hot season.

Mr. Singleton, of Bampton, is a great friend to winter vetches,

vetches, putting them in on a barley stubble in September, and beginning to feed them towards the end of May. Soils his horses on them in the stable, and raises by this means much manure. Sows wheat after them, which is as good as after a fallow, and often better; nor do they loosen the soil too much for wheat.

SECT. XI.—LENTILS.

THERE is a scattering of this pulse in Oxfordshire, but the quantity is not considerable. The account given me by the Rev. Mr. Filmer, of Lower Heyford, is, that the same tillage is given for them as for spring tares: sown the middle of March, about two bushels per acre. His have been made into hay, which is esteemed good for ewes at the time of yeaning: in a favourable season they produce one load and a half per acre. It is supposed for that purpose to be better than tare-hay; otherwise, inferior. The price of the seed, usually that of spring tares.

SECT. XII.—TURNIPS.

MR. Thos. Latham, of Clifton, has found that a new turnip, called the London white, succeeds better, and grows faster, than the Norfolk white.

Mrs. Latham sows trefoil with the crop of white corn, which is intended to be followed by turnips: it is fed by sheep in the spring, and then broken up for that crop. I saw some very excellent fields of turnips managed

naged in this manner, and without any other manure than that resulting from the sheep-feeding.

Mr. Davy, of Dorchester, sows trefoil for spring-food, and breaks it up very late in the spring for turnips; and he finds, that there are no finer turnips than those which he gains in this manner. All the circumstances which tend to shew that good crops may be gained without the usual number of ploughings, should be noted, that hereafter, like scattered rays of light, they may be brought to one focus. The land thus managed, cannot have been ploughed since the spring-sowing of the preceding year, and turnips are supposed to demand more ploughing than any crop; but here we find that the want of more tillage is made up by the roots of the trefoil, and the dung and urine of the sheep that eat it. The next time the land comes round for turnips, a regular fallow is given; not because the turnips want it, but that weeds may be the better destroyed by the variation of the tillage.

Mr. Davy, as soon as his drilled pease (which are well hoed) are cut, sows turnip-seed over the land broadcast, and has thus had turnips two feet round. I did not ask the question, but for this system he of course sows the earliest pea he can get.

Mr. Latham, of Clifton, is in the same husbandry; he sows the seed just before hoeing the pease the second time, in June, or the beginning of July: as soon as the pease are off, he hoes out the turnips; and these turnips are less subject to the fly than any other. It is never done but on land quite clean. He tried them among beans, but they did not succeed so well as with pease. 1. Turnips, 2. barley, 3. clover, 4. wheat, 5. pease and turnips, 6. barley, or oats.

Mr. Percey remarked on drilling turnips, that he had

had seen it on ridges; and the roots, in feeding off, rolled into the furrows, where the sheep trampled, and dunged, and staled on them, and in any wet season made them so dirty, that much waste ensued : he thinks the practice a very bad one. I must observe, that I have seen many scores of acres fed off, and never re-marked this, but rather the direct contrary; that the roots were eaten in a more cleanly manner than the broad-cast crops.

Mr. Shrubb, of Bensington, finds stubble-turnips more useful than others, as they stand the winter better, and are to be had when most wanted; for early ones, he sows the tankard.

The Bishop of Durham, at Mungwell, begins to plough for turnips in autumn, and works it well, and very fine, before sowing. To the credit of that very excellent farmer, Mr. Dean, of English, and upon difficult land, the finest crop of turnips I have seen this year, was one of tankard upon his farm; and on land so full of rough flints, that an ignorant person would wonder how any crop could grow. He dunged for wheat, and then folded for these turnips, ploughing the land four times, and sowing at Midsummer.

Messrs. Foster and Hairbottle, in the Henley district, cultivate this plant in the Northumberland system of one-bout ridges; the dung laid in the furrows, and co-vered by reversing the ridges; a single row is drilled on each. Mr. Hairbottle makes his ridges from 18 to 20 inches wide; Mr. Foster wider, in which I must think him right, as every inch of breadth renders the husbandry more effective, and is not, upon good land, any loss of crop even to 28 and 30. They have horse-hoes with three shares; the central one for hoei ng the furrow, and the others for cutting off a slice from each

ridge,

ridge, and, by a change, for throwing them back again to the plants. The rows are thinned, hoed, and hand-weeded by women. I have heard two objections to this system, and stated them to these gentlemen; first, that by the roots standing on the crown of the ridge, and the earth sloping from them, the sheep easily pull them into the furrows, where they are trampled, and receive the dung and urine of the animals. Second, that when the land is ploughed for barley, the manure is not spread about the land sufficiently, but left in lines perceptible in the crop. This is answered by first denying the fact: the turnips are held fast by the tap-roots, and the sheep scoop them out as well as in broad-cast crops; and as they are more disposed to remain in the furrows while eating, the roots are *less* soiled than in common crops. When tankard turnips are drilled, they will grow in various forms and positions, and some bending to the furrow; but this is not the effect of drilling, but the nature of the plant, and is found in broad-cast as well as in drilled crops. Respecting the second objection, they observe that their method is to plough the land across the ridges for barley, which, with harrowing, spreads the dung, so that the barley is as equal as after any broad-cast crop.

The farmers near Henley are very assiduous in ploughing stubbles *in* harvest, and before the wheat is removed for harrowing in seed for stubble-turnips; a practice that cannot be too much commended on such good sandy loam on gravel as that vale abounds in.

Mr. Tuckwell, at Cignet, finds that on the stonebrash soil of that vicinity, the earlier they are sown the better they are, especially if the land be poor; he has succeeded well from sowings in May.

Of late years, some farmers at Adderbury have got
into

into a new custom of laying on the manure for turnips in the preceding autumn, moving and mixing it with the soil by all the succeeding tillage; and they find their crops the better, and more sure, by this management.

" I would ask in what state can land be in, to promise better for a crop of barley, particularly if the summer be not too hot and dry? and in order to ensure as much as possible a crop of grass to succeed the barley crop, it will be necessary to sow the grass-seeds at the time of harrowing in the barley; as, in case the weather should prove dry after the barley is sown, it may be uncertain whether they may take so well, which ought not to be hazarded, as the crop of seeds is of as much consequence as a crop of corn. The ground being now cleared of the barley, the next scene presented to the eye will be a fine pasture of grass. Here the grower may force any part of his sheep-stock in as great a degree as the best pasture or meadow land in the kingdom: it should not, however, be depastured on later in the spring than Christmas, or Candlemas at farthest; as in case it should prove a dry spring, the hay-crop might be materially injured. The hay-crop being now secured, a crop of lattermath succeeds. Here, again, is another advantage to those who have but little pasture ground; and instead of having to search round the country for hired food to wean their lambs, at a considerable expense and inconvenience, they may do it at home much more to their profit and satisfaction. The second year's seeds will afford a sufficient pasture for ewes and lambs, or any other kind of stock the farmer chooses to depasture, till the time of breaking up for wheat, which usually begins six or eight weeks before Michaelmas, and is best to be finished
sowing

sowing at least a week before Michaelmas, always observing to have the land thoroughly wet at the time of sowing; which being done, you will generally have as good, and many times a better crop of wheat, than after a summer-fallow. The most profitable grain to succeed the wheat crop is oats, though a part of it is sometimes sown with beans or pease; but this is merely for the sake of having a little of those kinds of grain for the feeding of pigs, or for the horses when they come to eat green meat, but is seldom found so profitable as the oat crop.

" The six years being now completed, I shall briefly recapitulate the several advantages arising from this improved system, beginning, as before, with the turnips. The vast quantity of winter-food produced by a good crop of turnips, is of the utmost consequence where large flocks of sheep are kept, as is always the case upon stonebrash land, not only for the support of the store flock, but in fatting of sheep; and will keep them in a gradual state of improvement through the whole winter, which could not be effected by any other means without giving them corn. It hath also been found of the greatest utility in feeding of oxen or cows, to be given them with hay; and, as I before observed, the treading of the land in the winter is certainly a means of preventing its getting light and hollow, as is very commonly the case where a stonebrash soil is too frequently ploughed. The barley crop succeeding the turnips when the land is in such high condition, it is not uncommon to have as large a production of barley, as upon land of three times its value. The same cause may be assigned for a fine crop of grass for hay, which is succeeded by a good pasture for the sheep, till the time arrives for breaking up the land for wheat; and as

stonebrash

stonebrash land is often too light after a summer-fallow, it will be more likely to produce a greater crop of wheat when the land has been laid down two years, than after the expense and trouble of a summer's tillage. The next and last crop upon this system is the oats, which will be equally as productive as the last crop upon the strong three-crop land, if the summer be not too dry. I shall here state, that where land of this kind has been over-cropped, or where it is naturally very weak and poor, it is better to omit this last crop, and let the turnips succeed the wheat crop, at least for a round or two, till the land has a little recovered itself—all this the effect of a good turnip crop."—*R. Wills.*

" As I cultivate turnips in a way peculiar to myself, a relation of my practice should be known: the success is entirely to my satisfaction. At the moment of writing, my turnips are in full stock and feed. I prepare my land as early in June as can be done, fit for sowing; and having harrowed and rolled it, permit it to lay until the proper season for sowing, which I do not regulate by the time of the year, but the state of the ground, and the temperature of the air: the last season I began sowing in August. I consider the ground not fit to receive the seed until it has had sufficient moisture from the air, either by rain or attraction, which after some time will be the case, when the surface is perfectly dry; but I am never led by anxiety to sow too soon, and generally wait for rain. I consider the air fit for the process when it is mild, moist, and nearly stagnant.

" I generally have seed of my own growth, either the white or red tankard, or the Norfolk white, or procure seed of the same year's growth. The process is thus: with my Wiltshire eleven-share plough, fixed to the depth of three inches, I throw the land into furrows, **which**

which the plough finely pulverizes : the sower follows
the plough, sowing the seed broad-cast, the greater part
of which falls into the furrows, which, after the operation, is about two inches below the surface of the
land ; and after him, a light harrow covers the seed.
This process has been so successful with me, that I en-
tertain no doubt of its effect or utility. I felt not the
failure of last year. I am led to believe, that every
part of the process assists. To suffer the land to lay
after pulverization until it has imbibed a due proportion
of moisture, and is restored to the compactness necessary to fertility—to place the seed within the region of
moisture, and to place it at sufficient depth to give
strength to the radicle before the placenta are exposed
to the enemy—are all essential to the system. The vegetation is so rapid, that the depth of the seed produces
no delay in its appearance. The strength and vigour
of the plant bid defiance to the ravage of the fly, which
I have seen very busy on the ground, without injury
to the plant. I have never failed : I have indeed blanks
in the field, but those I attribute to other causes, viz.
the state of the common before severalty, alive with
rabbits, which had burrowed to great depths, and covered with furze, plots of which being impervious to
the sheep, the seeds are continually choking every crop
equally, and which for some time I must bear with patience. I sow no Swedes, having too many hares : my
neighbour this year was obliged to begin a crop of eight
acres in December, after two-thirds had been eaten by
them."—*T. Lowndes, Esq.*

SECT. XIII.—RAPE,

Is very little cultivated in Oxfordshire; but I found some on the rich red land north of Banbury. Mr. T. Wyat sows it mixed with tankard turnips, for weaning his lambs on: they begin the rape, and it teaches them to eat the turnips: this he finds very good food for them.

SECT. XIV.—CABBAGES.

Sir C. Willoughby does not know of any cabbages in Oxfordshire, except a small field of General Caillaud's.

Cabbages are uncommon in Oxfordshire; but Mr. Ashurst, of Waterstock, has planted some of the thousand-leaved for an experiment.

Mr. Salmon, of Hardwick, has a very high opinion of the cattle cabbage, and has cultivated them for some years; prefers August-sown plants, and gets much greater crops than by spring-sown plants; pricks them out at Michaelmas, and transplants the beginning of May: has them of a very large size; approves of them very greatly for sheep, generally fattening ones, and they do perfectly well on them; wintered 60 sheep on two acres, with the assistance of a few cut Swedes: the mixture of this food is very advantageous.

Mr. Thos. Payne, of Drayton, has this year a small piece of thousand-leaved cabbage on the red land of that district, which promises well, though not equal to the

soil,

soil, from having been sown too late, which was in April, and transplanted the 12th of June, at a yard square. Adjoining is a piece of green boorcole, which is inferior in luxuriance of growth. Of the large cattle cabbage he has been a cultivator these five years. He has two seasons of sowing—August for the early consumption, and February for crops eaten late : the former are pricked out in September or October, and planted where they are to remain, from the middle of May to the middle of June, according as rain falls. They are at their full growth the beginning of October, and last till after Christmas. The spring-sown plants will last till the middle of April. He finds them superior to all other plants for the quantity of stock they will carry—more even than Swedes. Gives them to sheep, cows, and fatting oxen, with the greatest success.

I saw upon his farm a very fine crop of autumn-sown ones, which had been planted this year (1807) after winter vetches, and had succeeded very well without manure on this red land. Mr. James Payne had also the goodness to shew me his crops on a farm about a mile from his brother's. The autumn-sown, after winter vetches dunged, were very fine indeed, and would now, if weighed, be found a great produce.

A part of the field on a fallow dunged, were also extremely fine : I guessed many of these to weigh above 20 pounds. He has them also on a distant farm, on clay, of great size. He saves seed from those which are stained with blotches, and veins of a pink red colour, being the hardest and heaviest. He is, from several years' experience, perfectly convinced of the great advantage of this crop; they are, for sheep and milch cows, better than turnips, and fatten cows admirably.

mirably. Nor do they exhaust the land, as he rarely gets less than six or seven quarters of barley, or 30 bushels of spring wheat, after them.

Mr. Warrener, of Bloxham, has tried them, and thinks they promise well. In 1806 he had six acres of the great cow cabbage, and five in 1807, which answered extremely well: eats them on the land with sheep, and gets better barley or oats than after any other preparation.

Mr. Creek, of Aston, seeing the cabbage culture of the tup-masters in Leicestershire, had some crops of them upon his own farm; but he is of opinion, that turnips exceed them in weight per acre.

SECT. XV.—RUTA BAGA, OR SWEDES.

Mr. Sarney, of Sounds, sowed four acres and a half the end of May, and thus kept 120 sheep nine weeks, from the 1st of March: this is 18*l.* at only 4*d.* per week, or 4*l.* per acre. The land had been laid down some time, broken up for oats, then turnips and barley, on which the Swedes.

Mr. Fane has had them, at Wormsley, for many years, and on a large scale: he approves much of them, and I was glad to find that they increase in the country: a party of excellent farmers I met at his house all spoke highly of them.

In Milton open-field I found a great extent of this most useful crop, well cultivated, and in great luxuriance; all broad-cast, and hoed out like common turnips. In a year in which common turnips have so generally

nerally failed, these crops must be invaluable. The soil, an exceedingly fine gravelly loam : capital land.

Mr. Davy, of Dorchester, has the highest opinion of this root, which he cultivates with much success : sows them early, hoes twice, and gains as heavy crops as of any other sorts; but much better, and more durable. He gets as good barley after them as after common turnips, eaten at the same time.

The Swedish turnip spreads very much about Baldon. Sir C. Willoughby saved seed of this plant, and giving it to his neighbours, was the means of considerably promoting the culture. Five years ago there was not any at Milton; and there are now, as I saw, considerable tracts of land, covered by fine crops of this excellent vegetable. That gentleman cultivates them regularly, and with much success. Half a crop of Swedes gives more nourishment than a whole crop of common turnips.

Mr. Thos. Latham, of Clifton, has cultivated this plant upon a large scale for three years, and found them to answer perfectly well. I viewed his crop of this year (1807), and found them remarkably fine, and very clean. He sows the end of May, or the beginning of June, dunging as for common turnips: they are all hoed twice, and well done. He does not find that his sheep break their teeth; but it is material to sow those of the yellow flesh. An acre would support half as many sheep again as an acre of common turnips; nor will they eat above half as much hay as they do with turnips. The barley after them is quite as good as after turnips; he has even found it equally as good, when eaten a fortnight after the others are gone. It is to be remembered, that the Dorchester district,

where

where Swedes generally succeed so well, contains much *red* sand.

Mr. James Welch, of Culham, has cultivated them two years. I viewed his crop of 1807, and found them very fine and clean, although sown after vetches.

The usual price of common turnips to feed upon the land with sheep, is 40*s.* per acre; but Mr. Latham values his Swedes this year (1807) at 4*l.*

Mr. Davy, of Dorchester, sows Swedes among his beans; when the crop is reaped, the poor people pull the stubble. I saw a large field in this state upon his farm; it was very clean; and though the plants were very backward, they promised to produce a good deal of sheep-food for the spring.

Mr. Bonner, of Bensington, cultivates and highly approves of this plant: sheep do much better, and fatten better on them, and fill themselves sooner with them, than on the common turnips. In feeding Swedes off, the sheep presently are satisfied, and lie down to sleep. For all turnips he ploughs deep—full six inches. He considers this plant as the best thing *ever found out.* Common turnips are so liable to rot, &c. that they cannot be depended on, and stock does far better on Swedes. The only objection to them is, their breaking the teeth of a regular flock; but for sheep to be sold off, that is of no consequence. He has found the barley after them to be full as good, or rather better, than after common turnips; but some think differently.

Mr. Shrubb, of Bensington, has an high opinion of Swedes, in utility at a most difficult season for supporting sheep: he thinks, however, that they have a tendency to draw the land more than common turnips;

and it is of ittle consequence to attempt them upon poor soils, for they demand very rich land.

Mr. Newton, of Crowmarsh, rarely has any Swedes, finding that his soil is too cold for them.

The Bishop of Durham cultivates them with success, and finds them very valuable: they are sown broadcast in June. I walked over 15 acres of them, which were very fine and clean ; and his common tankard turnips a good crop for the season. Of all sorts, his Lordship has this year (1807) 67 acres.

In the rich land of North Weston and Thame, there are very few Swedes : they have been tried, but were not liked, as Mr. Plaskett informed me. They have soils that would produce as fine crops as any in the county.

Mr. Dean, of English, has but few Swedes, as he finds that his land in general is not good enough for that plant.

The Rev. Mr. Filmer, of Heyford, cultivates them upon a large scale, and has a very high opinion of them. Common turnips are reckoned to winter six sheep per acre, if a middling crop; but Swedes do more, and at the most pinching season.

Mr. Blake, of Hampton Poyle, wintered, for two months, 20 cows on two acres of these roots.

Mr. Wyat, at Water Eaton, has a machine for slicing them, which he has found very useful.

Mr. Sotham has a very high opinion of Swedes, as they are of excellent use late in the spring: the roots are so numerous, that it makes him suspect their rather exhausting the land.

Mr. Pratt cultivates them with success, and has a very high opinion of them. He has found that sheep

will

will not eat common turnips while they can get at
Swedes. His barley is full as good, or rather better
after them, than after the others.

Mr. Kench, of Enstone, has a high opinion of Swedes:
he sows them the beginning of June, and gets as good
barley after them as after common turnips, unless they
are kept much later.

Mr. Rowland, of Water Eaton, cultivates Swedes
largely; and has so high an opinion of them, that he
is determined never to be without them for feeding
sheep.

Mr. Kimber, at Little Tew, has a very high opinion
of this plant, which he sows every year; but takes
much care, in saving seed, to keep none with fanged
roots—only with one root. Many take them up, and
preserve them in sheds, or other buildings, for feeding
fatting beasts, for which they answer well. If barley
be sown equally early as after common turnips, it is as
good; but if the Swedes are left to run for seed, they
draw the land.

Mr. Turner, near Burford, has a very high opinion
of Swedes, and gets as good barley after them as after
common turnips. In fatting his oxen that have been
worked, he finishes them with this root, and finds
that they answer the purpose of corn for finishing
such beasts.

Mr. Tuckwell, of Cignet, cultivates this plant largely
and successfully; but thinks that much more dung,
on stonebrash soils, is necessary to procure them than
is wanted for common turnips. His barley after them
is always much better than any other.

Mr. Tuckwell, of Cignet, and Mr. Pinnal, of Westall, Mr. Bagnal, and others in the neighbourhood of
Burford, are all great cultivators of this plant: they
have

have the highest opinion of it; but observe, that it demands great hoeing, to make amends for the early sowing preventing so much tillage as is given for common turnips. Their working oxen, and their flocks, all do remarkably well on this root, which is cut and given with chaff, and used till the end of May.

Mr. Edmonds has above 30 acres, and a great opinion of them as a resource, when nothing else is to be had, and particularly for spring feed: though they lose their leaves in the winter, what they produce in spring is very great. He gets as good barley as after common turnips, but they draw rather more: knows not what could be done without them.

Swedes are much cultivated about Witney. Mr. Coburn has this year a very fine crop: both he and Mr. Secker approve them highly, as they are of excellent use in the spring.

Swedes are much cultivated at Atterbury by all the good farmers there, and very highly approved by Messrs. John and Walter Wilson, Mr. Bellow, Mr. Garner, &c. Mr. Bellow has sowed them once or twice too late, and would now wish to have them sown the first week in June: they are hoed twice, and prove much more valuable than common turnips. I walked over a field of Mr. John Wilson's, part tankard and part Swedish; and had they been weighed, the Swedes would, I believe, have doubled the produce of the others. They get as good barley after them as after other turnips, nor will they ever be without them.

Mr. T. Wyat, of Hanwell, has for some years found these turnips so much superior, that he sows very few others: he keeps them till the end of May, and the beginning of June. He dungs for winter vetches, which are eaten off by horses tethered; then gives two

plough-

ploughings for Swedes, which are eaten the April and May following, and the land sown with turnips again. This is capital husbandry, except the single point of tethering ; but the land gets the urine. Mr. Thos. Payne, at Drayton, in the same district, cultivates them on a large scale, and with great success : I viewed with pleasure a large field of them, remarkably fine. He has carted off 20 loads per acre, and left a good sprinkling for sheep.

Mr. James Payne, neighbour and brother to the preceding gentleman, has also planted them largely, and has a very high opinion of them. I viewed his crops, and found them good and clean. He has found the application of them in fattening porkers to be the most profitable of all : they should run about as in common, only with as many Swedes as they can eat. Nine porkers paid him each 6*d.* per diem, for six weeks together, eating this root, which is a very remarkable fact, and highly valuable for the cultivators of this root to be acquainted with. His two brothers confirmed the fact, and had had great success in the same way themselves.

Mr. Salmon, of Hardwick, cultivates Swedes with much success : he has had them eight years. Sows the latter end of May, or beginning of June : his own seed, yellow flesh and single root, and the bulb quite round : hoes thrice, and gives them to calves, sheep, cows, and pigs, and are extremely useful to all : to pigs only the hulls, after other cattle. Uses them to the end of April. Sows spring wheat after, and gets as good crops as after any thing. They enrich so much by the stock kept, that barley would be all straw.

Mr. Davis, of Bloxham, remarks, that they must have manure, and therefore he is rather afraid of them ;

but

but has an high opinion of their utility. The fly ate
his crops for two years ; but as he was still in time for
turnips, the loss was nothing but the seed. Mr. Davis
does not manure for turnips ; as in that case the barley would be laid, and the seeds damaged, for he eats
all with sheep.

Upon viewing this intelligence concerning Swedish
turnips, I must remark, that this article of cultivation
makes a very distinguished figure, to the honour of the
agriculture of the county : it marks a vigour of exertion, in respect to this plant, which will be looked for
in vain in many others. It is completely introduced,
and under circumstances which promise a rapid extension : it appears from the notes, that the farmers feel a
thorough conviction of the importance attached to
them. The intelligence in this respect is almost uniform, and I examined such a multitude of crops in this
year so unfavourable to all turnips, and found the culture so well understood and so successfully practised,
that it is impossible, with any degree of candour, to
be sparing in commendation. It appears, from a variety of experiments, that May is the right time here
for sowing them : the season, however, extends through
the whole month of June. They are all sown broad-
cast, and hoed out from nine to twelve and fourteen
inches asunder. Good farmers give, all two, and
some three, hoeings : nor is the attention of hand-
weeding wanted, except in a few instances, in which
I have seen too much charlock scattered over some
crops in autumn. The use of them in feeding sheep is
well understood, and they are much relied on for the
late and difficult season in the spring. The improvement of slicing, and giving them in troughs to penned
sheep, much deserves attention ; and the great importance

portance of their application in the fattening of oxen, from stores preserved for that purpose, well deserves the imitation of many other counties.

The article of sowing winter tares upon a rich manuring, eating them off, and immediately sowing Swedes, is a practice that can scarcely be exceeded on good soils. The discovery that porkers are most profitably fattened upon them, would be alone worth a farming expedition to Oxfordshire: it highly merits to be well analyzed, by a series of experiments made expressly for that purpose; for should the same fact be established upon soils inferior in fertility to the red sands of the north of Oxfordshire, it would be a most important discovery indeed; and the nearly uniform intelligence, that the barley after them is to the full as good as that after common turnips, merits particular notice. The farmers who gave this intelligence were aware that any roots standing very late in the spring, must necessarily have a greater tendency to draw the land than if they were consumed at an earlier period. But it appears, from various of the minutes, that this circumstance is compensated by the sheep remaining so much longer on the land. When this crop is removed from the field, there is no more objection upon this account, than in the case of common turnips; but with the very superior advantage, that this root may be kept in any method with perfect safety, which every one knows cannot be done with turnips. All these, and various other circumstances noted in the minutes, prove sufficiently how well the plant is understood in this county; and is also a proof that the farmers of it are ready to attempt, and rapidly to extend, any article which shall appear truly to merit their attention.

SECT.

SECT. XVI.—CARROTS.

Mr. Salmon, of Hardwick, gave carrots to horses, and with much success, on a small scale; but means to extend them.

Mr. T. Payne, of Drayton, has cultivated carrots and parsnips on a small scale, and gave them to horses, cows, and sheep. He got just information enough from the trials, to be convinced of their admirable utility for all sorts of stock, and desisted from prosecuting the culture, merely from the great expense of hoeing and cleaning.

Mr. Warrener, at Bloxham, had an acre and a half of carrots in 1806, which produced a very good crop, and were of the greatest use in feeding eight and sometimes nine horses, during three months. This year he has the same quantity, but, from the dryness of the season, a less crop. Parsnips he has also tried; but they do not yield so great a crop as carrots, nor were they so valuable in the consumption. Mr. Thomas Payne, of Drayton, has been successful in both these roots. Mr. Warrener finds the carrots to be, of all food, the best for cows.

Mr. Harvey, at Broadwell-grove, near Burford, has cultivated carrots with success on stonebrash land: the culture is generally an article of such importance, that any experiment tending to prove that stonebrash will do for them, is valuable. I have no doubt of the fact.

All the sands in the vicinity of Dorchester and Nuneham would yield very fine carrots, and in general all dry, loose, friable loams, if eight inches deep. Finer land for them can no where be found than the sands
and

and dry loams of the red-land district ; and it is to be regretted that they are not more cultivated on all these soils.

SECT. XVII.—POTATOES.

Mr. Thos. Latham, of Clifton, has some acres of potatoes every year : he gives them raw to store pigs, and finds they answer very well. He sows wheat after them ; but in that case, thorough treading is necessary, and should not venture wheat without it. This year (1807) he intends to try them for cows.

Potatoes have been a good deal cultivated in Deddington-field by the poor people: the farmers dung and find the land; and the labourers plant, and clean, and take up: they divide the produce. It is done on the fallows, but considered as an injury to the wheat. The enclosure will do away all this business.

Mr. Turner, of Burford, is convinced that potatoes are an exhausting crop.

Mr. Welch, of Hanwell, has this year 15 acres of potatoes, cultivated by the poor people, who hire the land ; and then Mr. Welch puts in wheat, and gets good crops ; though wheat on this red land does not well upon fallow.

This root, upon the whole, is but little cultivated in Oxfordshire, compared with what it is in most of the counties with which I am best acquainted.

Mr. Lowndes has every year cultivated potatoes ; and has never had reason, when properly managed, to consider them an exhausting crop. He remarks, that a piece of land, comprehending an old brindle-road

and

and a decayed underwood, was grubbed, and planted
with potatoes three years in succession; afterwards
with turnips, then barley-seeds. Every year the fertility of the land has apparently increased: the barley
was too large a crop, and the seeds are very promising.
He always plants them in drills, made by a horse-hoe
with double mould-boards, at two feet rows; the drills
being dunged, the earth turned back, and the sets
dibbled in. Being duly horse-hoed, the same imple-
ment ploughs up the potatoes, keeping in employ
many pickers and carriers, and two carts. When in
sufficient quantities, he heaps them in what in Yorkshire are termed *potatoe pies*, which are thus formed:
the potatoes, dry and clean, are heaped in a conical
form, and covered with straw; then a trench is dug
round them, and the earth thrown over the straw, and
beat firm with a spade; after which they are slightly
thatched. In this manner they keep till the following
season for sowing.

" The steaming of potatoes, the use of them, and indeed their whole management, so well known and practised in other counties, cannot be too impressively
brought home to the Oxfordshire farmer; and this is
precisely the period of doing it, many of them having
shewn a willingness to cultivate them; but not knowing the use of them beyond their families or for market,
are indifferent about them. It is with pleasure I add,
that the only good potatoe I have ate this year, is the
produce of those sent me from the Board. I do not
grow them in sufficient quantities, either to set the example to my neighbours, or to inculcate the principles
of their application. It is true, I steam them for cattle,
and that by a method not devoid of utility; but being
more useful to a garden, dwelling-house, or manufactory,

tory, than to an agriculturist, I forbear specifying it. The fields of Dorchester and its vicinity, productive as their crops now are, might be trebled in productive food, if the potatoe culture were well established, and the propinquity of the Thames would ensure a market at their full value."—*Mr. Lowndes.*

SECT. XVIII.—CLOVER.

THIS plant has been so long in common cultivation, that its general properties have been for many years well known. In the case of such articles, when the soil, course of cropping, and prominent features of the husbandry are known, many others are necessarily concluded by a reader acquainted with the subject; and it is only in a few more peculiar variations that the curious inquirer can be gratified. The most important of these respecting clover, is the fact so severely felt, where it has been many years in culture, of dying away in the winter and spring, which the farmers describe, by saying their land is *sick of clover :* this renders a variation in the course necessary, that the recurrence of the plant so often as heretofore may be avoided.

Sir C. Willoughby sows clover twice in ten years, the land being new to it, and it never fails. He generally mows it twice for hay, and always finds his wheat the better for such mowings, and a better preparation than fallowing.

Mr. Newton, of Crowmarsh, for the sake of change, sows cow-grass; but it will not mow twice.

Mr. Dean, of English, has had white Dutch clover
for

for two years, and had as good wheat after it as in common.

Messrs. Foster and Hairbottle mow their clover once, and feed the second growth: the produce about a ton and a half.

Mr. Tuckwell, of Cignet, and many of his neighbours, do not mow more than one-third of the first growth of his first year's seeds, feeding two-thirds. The quantity of sainfoin they take care to possess, enables them to mow but only the small quantity of their seeds. The grass, thus kept for sheep, would afford a good bite for an ox the beginning of October.

Upon the gravelly loams of Henley, clover dies from repetition, so that they omit in their courses alternately. They substitute white Dutch and trefoil, pease, &c. but not ray-grass, as it exhausts the land; no good wheat to be got after it. Forty years ago nothing of this was perceived. They sow it now but once in two rounds.

I heard of some obscure notion, that sheep's dung will kill clover; but it was not well explained.

Mr. Weyland, at Wood Eaton, sows from 14 to 16lb. per acre: mows it twice for hay; some farmers once for hay and once for seed: one ton per acre of hay. The land here is not tired of it. The soil, various loams, and some stonebrash.

Mr. Sotham, of Stonesfield, finds that clover sometimes fails on stonebrash land; he therefore adds four pounds of white Dutch.

Mr. Kench, of Enstone, finds that clover will not stand, and is therefore forced to sow ray-grass: his working oxen are found useful in keeping it down in the spring. He mows the first year, and feeds the second, putting wheat in on one earth.

Great

Great Tew has been enclosed above 40 years, but the land is not tired of clover; and they sow sometimes trefoil and white Dutch, and at others from three to five bushels of ray-grass: yet the best farmers are convinced that ray-grass is bad for the land, and renders it foul. Two tons of clover are a great crop; one ton and a half middling; one bad. It is given to sheep in racks, while eating turnips.

The farmers around Caversfield consider clover, mown twice, to give better wheat than feeding; but Mr. Bullock doubts the truth of this opinion.

The Rev. Mr. Filmer, at Heyford, considers clover two years as liable to have more grass, but not to any degree to avoid the husbandry; and if it is so, to breast-plough the spots is considered as a remedy.

Atterbury has been enclosed many years, and Mr. Bellow finds, that clover does not succeed quite so well as it did from fifteen to twenty years ago. If it fails, he sows vetches, and eats them off with sheep. Mr. John Wilson, of the same place, in case of failure, ploughs, and drills early pease.

In 1806 Mr. Davis, of Bloxham, made an experiment of feeding his clover, &c. the first year, and mowing it the second, instead of the common management of mowing the first and feeding the second: the result, as far as he can perceive, is advantageous.

" The most material observation that strikes me as to the Oxfordshire mode of cultivating clover, is the species of manure, viz. the Newbury peat-ash, which comes into the Thames by the canal at Reading. About fourteen or fifteen bushels per acre are hand-sowed the latter end of February, or in March; its effect is immediately perceived, and it seldom fails of producing

a crop,

a crop, subject, however, to the observations incident
to the frequency in the cultivation of that plant. The
cost, with the carriage, may be estimated at 10*s.*
per acre, if you have the good fortune to obtain a to-
lerably fair measure, gradually decreasing in size."—
Mr. Lowndes.

SECT. XIX.—RAY-GRASS.

CLOVER and trefoil are perfectly well understood by
the very intelligent farmers from Culham to Dorchester : they sow no ray-grass, or at least very few of
them. When once they have sown it, they cannot
get rid of it for years, and conceive that wheat will
never be found so good where it is sown as where there
is none.

At Enstone, Westcot, Barton, &c. the most experienced men consider ray-grass as very far from an
ameliorating crop: an old and sensible labourer's expression was, *It is as bad and sucking as oats.*

Mr. Sotham considers ray-grass as making the very
worst of hay, unless cut young ; and wheat will always
be best where no ray is in the land. And though it is
known to be bad, yet on some land it must be had by
those farmers that are ignorant of better grasses.

Mr. Edmonds has seen a crop of wheat, with a part
where no ray, and the wheat evidently shewed better
just there. Sows no common ray-grass, nor any without a mixture of cocksfoot.

SECT. XX.——SAINFOIN.

I f I were called upon to name the circumstance in which the merit of the Oxfordshire husbandry is most conspicuous, I should be justified in quoting their culture of this grass upon all the soils that are proper for it, and to an extent, in proportion to their farms, that is highly meritorious.

Mr. Fane manures all his sainfoin with peat or coal ashes every year : 12 bushels of the latter, or 16 to 20 of the former. The benefit of the culture he finds to be immense : he has had 50 good waggon-loads of hay from 14 acres, and on an average, two loads an acre. But this produce depends very much on manuring ; for the fields which, with ashes, will give two loads, may not yield half a load without. This vast superiority has been long well understood. Mr. Fane's father thought his bailiff very extravagant in such constant buying of ashes, and objected to it. The bailiff ashed a field in the common manner, except about one acre, and there spread none. At crop time he took his master over the field ; and when Mr. Fane came to the spot not manured, he stopt his horse, and asked the man what could be the matter with that bit? When the reason was explained, the conviction was complete ; and the bailiff had orders to proceed in buying all the ashes he wanted.

It is agreed by all, that it is not so productive, nor so durable, at present, as it was in times past ; and attributed to the repetition on the same land : it once lasted fifteen and to twenty years ; now ten or twelve, and some not more than eight. They break it up for,

1. oats,

1. oats, 2. turnips, 3. barley, 4. pease, vetches, clover, or trefoil; 5. wheat, 6. turnips, 7. barley, and sainfoin again. Upon fresh grubbed woodlands, it does not succeed so well as on other land : the soil left in too hollow a state. But beech being as natural to chalk-hills as sainfoin, *query*, If they do not rob each other ? Formerly, the culture was confined to the hills ; but it has now got down into the dry vales. There are few districts in which it is so much cultivated as here : full one-seventh of all the arable land is under this grass.

Mr. Simmons, of Stokenchurch, has fed sainfoin with sheep, and with much success. It is well known, that sheep eating the first spring shoot of this grass is highly injurious to it, as they are at that season apt to eat into the crown of the root, and greatly weaken, if not destroy, the plant. But Mr. Simmons ate the whole growth, not turning in till the crop was well advanced ; and he found that, in this case, the plant was not damaged.

It is ascertained on the Chiltern hills, that if cows are turned into sainfoin rouen before some slight frosts come, the cream is bad.

"Certain farmers, much esteemed, have laid down sainfoin on very light land, with the last crop of corn : the first year it is not so luxuriant ; the second year, nearly equal ; the third, and afterwards, equal, and lasts full as long : always couch in the last crop, and will last six or seven years, and by this means give more ashes."—*Mr. Edmonds.*

Mr. Newton, of Crowmarsh, finds that sainfoin is of such consequence to good husbandry in this country, that he has 100 acres of it. He spreads from ten to twelve bushels of peat-ash per acre every year over the

whole

whole of it, as well as upon all his clover. It would last longer, but he does not approve of leaving it for more than six or seven years : and although it is subject to grubs and red-worms, he does not break it up by paring and burning, thinking the land too stony for that operation. He dungs it the end of July, or the beginning of August ; ploughs it in a month afterwards, and in three weeks sows wheat, upon which he pens his sheep. The carting and the penning firms the surface, and prevents the depredations of the worm : there is no object more important than to keep these loose soils tight and firm.

The Bishop of Durham, at Mungwell, much esteems this crop ; he always mows it : the produce, two tons and a half per acre ; but ten bushels per acre of peat-ashes are sown upon it every year. Upon the hills, it lasts eight or nine years ; but on the lower grounds, ten years. He breaks up by ploughing, and gets very good crops after it, so as to prove that the land is the better for it. The crops are not often attacked by the wire-worm, the sheep-pen seeming to prevent it. The same land should not be sown again. The hay is used for horses, but not for sheep, as it is too stalky for them ; a circumstance not to be prevented by early mowing, as it is here thought that it should be cut only when in full bloom.

Mr. Kelsey, of Whitchurch, has 60 or 70 acres of sainfoin, the whole of which is mown every year, except one piece, which having been sown with wheat in October, suffered by the frost, the season being alternately fine and frosty. In order to improve it, he has seeded it every year for four years, which has answered in a considerable degree. The crops of hay on the worst hills, about a ton per acre ; and on those rather better,

OXFORD.] from

from one and a quarter to one and a half: the price, generally from 4*l.* to four guineas. He gives it to sheep.

Mr. Gardiner, of the same place, has an high opinion of it: it lasts with him ten years, but is never broken up by paring and burning; yet the worm often commits depredations on the oats sown on it.

There are considerable tracts of dry hard gravel, with a good red sand surface, in the Dorchester district, upon which it might be conceived that sainfoin would do well; but it has been tried by Mr. Davy, and by Mr. Latham, and it was not found to answer.

Mr. Bonner, of Bensington, thinks sainfoin, on poor thin land, the most profitable of all crops; but on gravelly loam, tillage more beneficial.

Mr. Dean, of English, has 40 acres of sainfoin: it lasts five years. When broken up, it is by ploughing, and never by paring and burning. The first crop he sows is oats; he then fallows for turnips, but it is not kind for turnips. The oats are good; but it is by means of sheep-treading, and heavy rolling with iron rollers.

He does not give sainfoin-hay to sheep, as the lop-grass in it fouls the land exceedingly: uses clover-hay instead of it. Much sainfoin is sown at Maple Durham: five bushels of seed per acre. The after-grass is fed by cows, and by lambs at times. Peat-ash is very generally sown on it. It now lasts but seven or eight years, but formerly longer. They plough it up for wheat, and get good crops: some of the best this year was gained in this manner; but much folding with sheep upon it, before and after sowing.

Mr. Turner, of Burford, does not think that it is necessary to have land very clean for this plant: he

has

has twice had it upon foul land, with success equal to the cleanest. At the time of Burford enclosure, a piece of 15 or 16 acres, the lands of which ran in different directions, and, when open-field, in the occupation of three farmers, two excellent ones, and the third a very slovenly cultivator; one-third of the piece was very foul, and two-thirds quite clean. The land was ploughed across once, and sown with barley—with it sainfoin, and four pounds per acre of trefoil. The first year of the sainfoin, the crop on the clean land was pretty fair, and that on the foul piece not half a crop; the second year, the products were more equal; the third year, the one as good as the other. It lasted eight years longer, and that upon the foul land incomparably the best.

Mr. Turner would never on any account break up sainfoin without paring and burning: he has seen oats sown in such a case on merely ploughing, and the crop entirely eaten up by the grub. In cultivating this excellent plant he has always found, that seeding a crop improves it.

Mr. Tuckwell, of Cignet, one-seventh of whose farm is always under sainfoin, does not credit a notion that has spread in many places, that after land has been under sainfoin, it must not be sown again with it for many years; and a case which occurred to himself confirms him in the opinion. He had a piece of two acres, which, in altering a field, was thrown to another going to be laid down with this grass. That two acres had been under sainfoin many years, and had been broken up only two; but that the field might all be under one crop, Mr. Tuckwell sowed it with the rest, and it did just as well, and yielded as good crops, such as two tons per acre. But great crops are only to be gained

by

by means of coal-ashes. He sows three bushels of seed **per**
acre, also from one pound and a half to two pounds **of**
trefoil. This is only for an addition to the first crop,
and they never see any more of it. Some sow ray-grass
with it, but he disapproves much of this practice.
Many sow four bushels per acre of sainfoin. Dung is
known to be pernicious to this crop, filling the land
with ray-grass, and other plants which damage the
crop. Mr. Mewix, of Burford, tried it, and ruined
his field.

Mr. Tuckwell remarked an opinion not uncommon,
that mowing sainfoin young hurts it ; but he has experienced the contrary, in mowing it very early for soiling horses and oxen, without the smallest prejudice.
He has also found, that to feed sainfoin with sheep for
a year before breaking up, is very advantageous in increasing the ashes, by thickening the turf to be pared.

Mr. Pinnal, of Westal, has more than one-sixth of
his farm under sainfoin. He breaks all up by paring
and burning, and would on no condition whatever do
otherwise. Upon that operation sows turnips ; and
seeds never take so well as with barley following such
crop. Lays down again with sainfoin in seven, eight,
or ten years, as may be wanted ; the notion that a
longer time is necessary having been proved erroneous.

The first year of sainfoin is considered as giving better hay than any growth afterwards.

Sainfoin is sown after turnips at Stonesfield, from
three bushels and a half to four bushels of seed per acre,
with four pounds of trefoil. It is mown every year,
yielding after the first, one ton and a half per acre : the
after-grass is applied to feeding lambs, but not longer
than the middle of October. The last three years, Mr.
Botham has cut 60 tons from 30 acres : but if it yields
in

In this manner for two or three years, it is apt not to last so long as it otherwise would do. The hay is of such consequence, that they depend upon it, and do not mow their seeds at all : it sells from 4*l.* to 4*l.* 10*s.* per ton, being carried two miles ; and he thinks sheep will pay 3*l.* 10*s.* per ton for it. When they break it up, it is all by paring and burning; rice-baulking it first, and then a clean earth for turnips. Barley follows, and the seeds sown with it are always capital. The land should not be sown again for fourteen years ; if oftener, it lasts only three or four years, and it would be better if only once in 25 years. The after-grass is the best food for lambs, and it never scours them. Mowing the crop for seed improves it; and it never ought, in the opinion of Mr. Sotham, to be cut early, as it bleeds and is hurt by it.

The Duke of Marlborough has, at Blenheim, sainfoin to a large breadth, and finds that it yields the very best hay. Mr. Palmer fattens the oxen that have been worked, and many others, entirely on it after they are taken into the stalls, and they do perfectly well on it. When broken up, it is by paring and burning for turnips; and I saw some very fine crops thus gained, as well of the common sort as of Swedes.

Much sainfoin on all the stonebrash lands from Chipping Norton, in every direction. It lasts ten years, and where it has not been sown before, would last fifteen ; but Mr. Kimber, of Little Tew, thinks it more profitable than to break it up by paring and burning for turnips, and sow down other land. So long as there is fresh land to be had for it, he would not sow it again till 20 years are expired ; but if the farm has all had it, this must be dispensed with ; for the object is so valuable, that some must be had. He never ploughs

it

it without paring and burning. Fresh grubbed furze land, though on stonebrash, will not do for it, from being, as they conceive, too hollow; and a very bad white grass is sure to thicken and damage it greatly.

It has been found, in many trials around Chipping Norton, that sainfoin will not succeed upon old heath land, on which furze was the spontaneous growth: it presently goes off, and is lost.

At Enstone and the vicinity, it lasts ten or twelve years; but if the land be wet from springs, it goes soon: on a true stonebrash it will last longer. They value it highly: the culture increases. When they break it up, some plough it for oats; but this is reckoned a very bad way, as the red-worm often eats up the crop: but the best farmers pare and burn for turnips, after barley, &c. The land is in a very hollow state, and wants the treading of sheep in eating off the turnips.

They do not sow any ashes on sainfoin about the Heyfords, &c. yet their average crop of hay is not less thon 30 cwt. : the common proportion of this grass is about 30 acres on a farm of 400. The principal application of the hay is for sheep and cart-horses: for the former, while at turnips, it is superior to any other, but highly valuable in every use.

Mr. Foster, on his stonebrash soil at Bignall, has much sainfoin, which he values highly. It lasts twelve years ; breaks up by ploughing once for oats, then turnips.

Mr. Bullock, at Caversfield, observes, that this grass is one great staple of their stonebrash soil, and exceedingly useful. On new grounds it lasts fourteen or fifteen years, but after being sown thrice, only seven or eight.

eight. They break it up for oats, then turnips : some
pare and burn, but it is not common.

Near Witney, it lasts now not more than eight or
nine years ; formerly longer. They break it up by
paring and burning for turnips, which are always
good. The grass is all mown, and the rouen eaten by
lambs : the hay, and all other hay, given to sheep.

Mr. Secker has one-seventh of his arable under this
grass, which, upon poor stonebrash, he thinks more
profitable than the common tillage course ; but not so
on better soils.

As far as Mr. Davis's observation goes, he does not
think that it will succeed on red-land—on stonebrash,
incomparable. *Perhaps* the want of leases may be the
reason of the farmers' breaking it up sooner than they
would do, were they in that respect differently situated.

" As I have not before said any thing of sainfoin, I
shall here give my opinion of its utility, which is certainly very great where there is not a good proportion
of pasture or meadow land annexed to the arable, and
is the best substitute now in use, the product being as
large, and, if cut young, equal in quality for almost
any purpose, to hay cut off the best meadow or pasture
land : the lattermath is also very excellent for weaning
lambs, for milking cows, and almost any other kind
of stock that require to be kept well. The plant will
generally continue in perfection for four or five years ;
and if the land is fresh, and natural for this kind of
seed, it will go as far as six or seven years, when it
will begin to decline. Care should be taken not to feed
it too hard with sheep in the autumn, as that will be
apt to damage the plant ; but it is usual to depasture
it with sheep one year, at least, before it is broke up.
The

The best method of breaking up old sainfoin, **or old down** land, is by paring and burning, which should be constantly followed by a crop of turnips; the grand object in breaking up old sainfoin, or old down land, being to get the land in as firm a state as possible before it is sown with corn, which cannot be done by any other means so well as by turnips, and it has been found by experience the best means of procuring a crop of them. The objections to paring and burning, in my opinion, are futile, it having been in practice upon the Cotswold hills of Oxfordshire and Gloucestershire for time immemorial, and no visible inconvenience has appeared; unless where it has been repeated every ten, fifteen, or twenty years, which is certainly too often. I have one farm in particular, of the lightest stonebrash soil, where it has been in constant practice for near 40 years, and is perhaps at this time in the highest state of cultivation, of any land of the same nature in the county of Oxford."—*R. Wills.*

This noble plant is probably cultivated upon every soil in Oxfordshire which is fit for it ; it is one of the most distinguishing features of the husbandry of the county : the cultivation is well understood, and very successfully practised ; and though in some districts it is broken up sooner than perhaps it ought to be, yet this is probably the result, not so much of a want of understanding the value of the plant, as of the want of leases. After a certain number of years' growth, a considerable benefit is made by breaking it up ; and a tenant, apprehensive of quitting his farm, will not be very ready to forego the early acquisition of such advantage, lest the benefit should, if he had more patience, accrue to another : it is thus that the want of leases is felt in the regular standing cultivation of the farm. The

The preparation for the plant by turnips—the quantity of seed sown—the attention to keep sheep from it at improper seasons for feeding—the very high idea they have of the value of the hay, and the conviction they feel of the propriety of breaking it up by paring and burning, and this without taking any extraordinary number of corn crops in consequence of it—altogether prove the cultivation of it to be perfectly well understood in this county. In the Burford district, the having one-seventh of the arable land under sainfoin, is a scale of applying this crop that does great honour to the enlightened farmers of that vicinity. Sainfoin is pretty well understood in some parts of Norfolk ; but where are we to go in that county to find a seventh of the arable under this plant ?

SECT. XXI.—LUCERN.

Mr. Freeman, of Fawley-court, has three acres of lucern, which, though very much-neglected, yielded this year two good cuttings ; sufficient to prove, that the soil is very well adapted to this plant.

Mr. Sotham informs me, that it has been found that stonebrash soil will yield lucern. Mr. Price had it at Wootton : he mowed it three times a year, and it gave a great burthen ; but it was manured every year.

Mr. Thomas Payne, at Drayton, shewed me, on a border, the remains of a piece of lucern, which were sufficient to prove, that the red-land district is well adapted to the culture of this plant, of which, indeed, no doubt could be entertained.

There is none in the part of the county south of Oxford ;

ford; but I was assured by a sensible man, that he had known horses eat the first growth of it, but would not touch the second. Any person engaged in such inquiries, will often have occasion to remark assertions that make him stare.

SECT. XXII.—CHICORY.

MR. Fane, at Wormsley, drilled it in a field of very poor land, and found it of such use in feeding cows, sheep, and pigs, that he has a very high opinion of it. As clover dies, and a tendency in sainfoin to shorten its duration, any grass that is suitable to these hills becomes an object of consequence to every farmer.

SECT. XXIII.—BURNET.

MR. Turner, near Burford, had 15 acres; but it did not answer like sainfoin.

Mr. Large, of Bradwell, also tried it; but I could not learn with what success, as, unfortunately, he was absent when I called at his house.

ASTRAGALUS CICER.

Of all the plants in the Botanic Garden at Oxford (viewed in September, through the very friendly and obliging attention of Dr. Williams, the professor), the two which were most promising were, the *Bunias Orientalis*

entalis (which I have had for some time in cultivation at Bradfield), and the *Astragaلus cicer.*

RHUBARB.

Mr. Thomas Payne, of Drayton, in the rich red-land district, has rhubarb four years old, three years, two years, and one year : it is planted in squares of four or five feet, and kept clean. He conceives that some of the older roots weigh half a hundred weight, and he has been offered 2*d.* a pound for it green ; thinks he can get 3*d.* : if so, no crop can pay better.

SECT. XXIV.—HOPS.

Sir C. Willoughby has had a hop plantation of five acres for 27 years, that have produced him from no crop at all, to 18 cwt. per acre ; and the price has been almost as variable as the produce : when he had 18 cwt. they sold at only 3*l.* 3*s.* per cwt. ; last year (1806) he had 4 cwt. and the price was 7*l.* 7*s.* : one year, 14*l.* 14*s.* His main object in this cultivation has been a home market for his ash-poles, of which he has a plantation, which thus pays him 3*l.* per acre per annum. The soil, a sandy loam. His expenses have been, every thing included, 30*l.* per acre. All labour he pays by the day, except picking, for which $1\frac{1}{2}d.$ per bushel. Average produce, 5 cwt. The means of renewing the plantation is, to grub up a space, and plant an equal space every year. In order that the culture may not rob the farm, Sir Christopher buys rags from London for a partial manuring of the hills ; this he has

done

done to the amount of 40*l.* in one year, but he thinks they have a tendency to cause the mildew. Some stable dung, but not nearly so much as of rags. This year no hops.

Mr. Hayward, of Watlington, who is a considerable brewer, uses his hops as manure, and with great effect; they are so strong a dressing, that if more of them be spread than of ashes (15 or 20 bushels per acre), the corn will be laid from rankness.

———◆———

SECT. XXV.—HEMP.

It is not 40 years ago since every cottage at Baldon had a plot of hemp, and all manufactured into linen for their own consumption, selling what they could spare: at present, no such thing; the last was given up about six years ago.

———◆———

SECT. XXVI.—FLAX.

Eight or nine years past, there was a considerable quantity of flax raised at Water Eaton, Hampton, and Yardington, on boggy land; and good wheat got after it by Mr. Cocks, &c.; but at present there is none. A very singular husbandry, however, in this vicinity, has been the culture of this plant for the object of seed, for the sole purpose of fattening bullocks: the high price of linseed-cake occasioned this management, which answered well. The flax was watered, and dressed as in the common way; but the object of the cultivation was the seed for live stock.

CHAP.

CHAP. VIII.

GRASS LAND.

————

SECT. I.—MEADOWS.

" THE open-field meadows are often situated a considerable distance from the villages, and besides, generally lying in very narrow slips and parcels, are frequently even in lots changeable every year. These common meadows seldom receive any assistance of manure, because the arable lands consume the whole of the manure in the preparation for the wheat crop. There is also another reason why the common meadows are neglected in this respect, because the after-feed is the property of all the occupiers in the parish at large. And to this circumstance may be attributed another ill consequence, which is, that in meadows where there is no fixed time for turning in the cattle, the grass is suffered to stand so long, in order to obtain as much bulk as possible, that though there may be a considerable *quantity*, yet the *quality* is very inferior, having stood till the nutritive juices are dried away. The greater part of these meadows are near rivers, and are situated so low, as to be overflowed occasionally after hasty rains, and now and then even in hay time, insomuch, that the crops are either entirely swept away, or so greatly damaged as to be of little value, except for littering the farm-yards. This overflowing is accounted to improve the meadows, when the water does not con-

tinue

tinue long enough to chill the grasses, so much as to destroy the more valuable sorts, and to cause a succession of aquatic plants, of inferior value, to take their places.

" And though there is a variety of opinions, whether the improvement is greater from foul or clear water, yet the most probable opinion is, that it is the spirit which the land imbibes from the water that brings on a fermentation, and promotes vegetation; and therefore, when the water has become foul by running over poor soils, that spirit is in some measure gone, unless in the case of its having passed over arable or other lands which have been manured, and bringing with it a certain portion of such manures, which settles and remains on the land. But where the current is less rapid, and the water becomes more stagnant, the greatest injury takes place; and therefore, the embanking of this kind of meadow land, so that the water might be admitted or kept out at pleasure, and keeping the rivers and principal water-courses properly cleansed, would be a very great improvement*."

" A very large tract of valuable meadow land in Oxfordshire and Northamptonshire, on the banks of the river Cherwell, has been much injured, and in many places spoilt, by a navigable canal made immediately above its level, and, from Banbury to Oxford, very ill executed: the extent from the first-mentioned place to where the canal leaves the meadows, is about 20 miles. It is rendered extremely boggy by the continual oozing of the water through its banks; and, in lieu of meadow-grasses of the best quality, with which it before abounded, is now over-run by *caltha palustris*

---

* Original Report.

(marsh

(marsh marigold), and other aquatic plants. This evil, if not remedied, will increase daily."—*MS. Annot.*

At Water Eaton is the best grass land in the county; it is under dairies. I examined these fine meadows, and found their appearance and stock to justify the character I had heard of them; but they are subject to summer floods, which sometimes do much mischief, insomuch, that Mr. Rowland has had five hundred pounds' worth of hay lost, or greatly damaged, in a single season. However, these meadows are said to have lett at 40*s.* an acre, 40 years ago; but every one knows, that meadows have not increased in rent equally with various other lands: their produce has been rivalled by most of the modern improvements, and covering the hills of calcareous countries with sainfoin has given a new value to them, somewhat at the expense of all the meadows of a district: many of these are, however, said to lett at 3*l.* per acre.

At North Weston, in the rich district of Thame, they mow their meadows twice a-year, without any manure being spread on them; and this practice is deemed prejudicial. These meadows are subject to summer floods.

Mr. Tuckwell, of Cignet, like his neighbours, has very little permanent grass of any kind; but being sensible of the great utility of kept grass, he preserves a part of his seeds from the beginning of August, to feed with sheep in the spring, and finds the benefit of the practice very great.

" *Herbage.*—The plants that constitute meadow land, differ according to the variety of soils; some being more the native produce of clay—others of loamy, sandy, or moory grounds; and again, they
vary

vary according to the dryness or wetness of the soil.
But the following account of two meadows, examined
at South Leigh, contains the most predominant plants
and grasses that are found in the meads of this county,
with this difference, that in various soils, some of the
kinds are more abundant than others.

"The first was a meadow, which was sometimes
overflowed, wherein the grasses were,

1. *Alopecurus pratensis,* Meadow foxtail.
2. *Poa trivialis,*   Rough-stalk meadow-grass.
3. *Festuca fluitans,*   Floating fescue.
4. *Poa pratensis,*   Smooth-stalk meadow-grass.
17. *Carex panicea,*   Pink-headed sedge-grass.
18. *Carex gracilis,*   Slender sedge-grass.

"The leguminous plants were,

5. *Trifolium repens,*   Dutch, or white clover.
6. *Trifolium fragriferum,* Strawberry-headed trefoil.
7. *Vicia cracca,*   Bush vetch.
14. *Pucedanum silaus,*   Meadow sulphur-wort.

"The general meadow plants were,

8. *Plantago lanceolata,*   Narrow-leaved plantain.
9. *Sanguisorba officinalis,* Greater burnet.
10. *Leontodon autumnale,*   Autumnal dandelion.
11. *Spiræa ulmaria,*   Meadow sweet.
12. *Thalictrum flavum,*   Meadow rue.
13. *Scabiosa succisa,*   Devil's bit.
15. *Prunella vulgaris,*   Self-heal.
16. *Rhiananthus crista-galli,* } Common rattle.

"By numbering the above it is intended to shew, in
what estimation the herbage may be held of each, beginning with No. 1; so that those pastures which

                                            abound

abound most with the higher numbers, may be considered as of little value. For instance: No. 14 is bad and hot to the taste, and No. 17 and No. 18 are of little value, being the effect of neglect of draining and manuring, by which methods they are got rid of.

" In another meadow, which was *moist*, but *not flooded*, the grasses were,

1. *Alopecurus pratensis,* Meadow foxtail.
2. *Poa trivialis,* Rough-stalk meadow-grass.
3. *Poa pratensis,* Smooth-stalk meadow-grass.
4. *Festuca pratensis,* Meadow fescue.

" The leguminous plants were,

5. *Trifolium repens,* Dutch, or white clover.
6. *Trifolium pratense,* Meadow trefoil.
7. *Lathyrus pratensis,* Meadow vetch.
8. *Lotus corniculatus,* Bird's-foot trefoil.
13. *Pucedanum silaus,* Meadow sulphur-wort.
14. *Ranunculus repens,* Meadow crowfoot.

" The general meadow plants were,

9. *Plantago lanceolata,* Narrow-leaved plantain.
10. *Sanguisorba officinalis,* Burnet (purple-headed)*.
11. *Leontodon taraxicum,* Dandelion.
12. *Spiræa ulmaria,* Meadow sweet.

" The foregoing account of herbage may be considered as comprehending the most general only, amongst the numberless plants and grasses with which meadow land and pastures abound. But there are in this county two capital collections of them; one at North Aston,

---

* " This is the burnet on *wet* land, being different from the *poterium*, or burnet on *dry* land, which bears a green head, and is the species that is cultivated."

OXFORD.] the

the other in the Botanical Garden at Oxford; and it
may be considered as an acquisition to the public, that
the present Professor of Botany has attended much
to the British grasses, and from whose superior knowledge of the qualities of plants and grasses, much may
be expected.

" At North Aston are to be seen above a hundred
specimens of different grasses, together with about
twelve hundred plants. The garden for the aquatics is
admirably well constructed, and all are kept in excellent order. One of these plants is particularly worthy
of notice, namely, the *Festuca ovina vivipera*, brought
from the top of Snowdon; it is of a very nutritious
quality for sheep, and is said to abound in Spain, and
to contribute in producing the fine wool of that country.

" As it is very material and desirable to bring pasturage to perfection on arable lands, I shall beg leave,
before I quit the subject of grasses, to mention a new
species of rye-grass; and though to do this I must pass
a little beyond the bounds of this county, I trust the
digression will be excused on account of the importance
of the occasion, as I am fully convinced, from repeated
observations at various seasons, that the grass in question has a manifest superiority over the common sorts.
It should also be noted, that the spot where I viewed
this grass, at Northleach, in Gloucestershire, is of that
shallow stonebrash kind, with which a considerable
tract of Oxfordshire, about Burford, abounds.

" The excellencies peculiar to this species of rye-
grass are the following :

" 1st, That in the autumn, when the other sorts are
become of a russet hue, withered and decayed, and
produce

produce little feed, this is luxuriant and growing, the tufts thereof spreading over more than twice the space of ground that the common sort does.

" 2dly, That it will remain in the ground for seven or eight years, or more, depending on the quality of the land; whereas the other sort will not continue above one or two years, which is too short a time to give sufficient rest to the poorer sorts of land.

" 3dly, A particular advantage arises by its being hained up about Michaelmas, or before, whereby it will grow at all open times during the winter, and produce a valuable pasturage for the ewes and lambs in the spring of the year, when the turnips are gone.

" The merit of discovering and cultivating this grass is to be attributed to that ingenious and intelligent farmer at Northleach, whose attention to the breed of sheep has made his name so well known to the public. It first attracted his notice by continuing amongst sainfoin which had stood seven years ; the seed was therefore first selected from this sort, and from time to time multiplied, till the cultivator has been able to accommodate many of his friends, and the public (who have now found out its value), with considerable quantities of the seed : but not so much as has been lately required, 50 quarters having been sent for in one order in 1792, when the whole that was raised that year did not exceed 16 quarters. It therefore seems to be of public utility, that this valuable seed should be disposed of in smaller quantities, in order that it may be more universally dispersed ; and it has been the advice of several gentlemen, well-wishers to improvements in agriculture, to raise the price to 10s. 6d. per bushel, including new sacks to send it in, which would have the good effect of causing those farmers to save seed
who

who now feed it off under an expectation of procuring
more seed from the same place, at the original price of
5s. per bushel ; and the cultivator is undoubtedly entitled to such an advance, for his attention to the public
interest, in selecting, cultivating, and preserving the
seed.   One bushel is sufficient to sow an acre ; and as
the plant comes up weak the first year, it is advisable
to sow it amongst corn, in order that the weeds may
not get the better of it ; or it may be sown amongst
turnips, hoeing it in.   But if the land is intended to
be a permanent greensward, a mixture of the *Dac-
tylis glomerata*, or rough cocksfoot-grass, would
prove beneficial, the seed of which is now selecting
also by the same person *."

" I have to remark, that whether grasses or other
herbage, they demand to be supplanted by a wholesome herbage. It is very questionable if the sheep ever
find an inclination to an unwholesome herbage, but
from a degree of necessity, impelled by an unsatisfied
appetite.

" I once lost the least ailing and most sprightly horse
I ever possessed, by his cropping, early in the spring,
the new shoots of the monkshood.

" It may not be unuseful to commemorate, that this
horse escaped out of his stable one night into a little
compartment where grew various plants, all which he
cropt close to the ground, save this monkshood : I
hence mistakingly reasoned, that his instinct of its
baneful quality would be to him a constant preservation. Several evenings after this he broke out again,
cropt the monkshood, and in a few hours was seized
with pains that made him outrageous. At length,

---

* Original Report.

oils,

oils, &c. were urged down his gullet : he languished a few days, and died. Still the grass alluded to, the bane of sheep, may not be detrimental in its native quality; but may, at a length of time, be a receptacle of the *ova* of pernicious insects, which afterwards hatch into a vermicular state, and insinuate themselves into the vital seat of animal life*. This, I conceive, is not a new theory; but whether the rank green herbage is rendered malignant by time, or become the receptacle of pernicious seeds (eggs), either circumstance demands an early cropping, manuring, or eradication; hence the certain salubrity of ditching the soil, and every species of draining."—*MS. Annot. Mr. Wagstaffe, of Banbury.*

" There can be no doubt that divers meadows are not productive of suitable herbage, through that neglect of improvement which, in certain periods of the year, is practicable on all, and in course, those subjected to inundation : for want of this improvement, weeds sometimes occupy an eighth part, and sometimes nearly a fourth part, of those most profitable grounds : when they may be in the hands of those who are determined to reduce, to eradicate those useless, and, in an agricultural sense, pernicious weeds; such are the meadow sweet, rue, crowfoot, &c. The carexes, indeed, multiply greatly; but the wholesome will, generally speaking, be paramount over them : but those I have mentioned, particularly the *Spiræa ulmaria*, will domineer over every grass and trefoil at length, to their exclusion from the scite they possessed.

---

* " I conceive the botts in the stomach of horses have a similar origin, and which are the source of diseases of different names, that often destroy that useful quadruped."

" This

" This gentleman has judiciously remarked, that by draining and manuring they may be got rid of. Draining alone will weaken, and may destroy them, and manuring encourage the grasses to a predominance over their overwhelming adversaries ; but a short and certain cure of the perennial weeds, of which those I have mentioned are a part, is by the spade or the plough : first premising, that the ditches should be cleansed, and their contents applied to the upland arable; then taking off a crop of hay, and a while feeding the aftermath. No doubt but the land is now well drained, perhaps the ditches are dry, and the ploughs can then traverse to their edge ; and where the plough cannot safely go, the spade more effectually answers, as on the edges of ditches and unopened rivulets. The meadow sweet and rue predominate, and to which may be added the willow herbs (*Epilobrium palustre* and *hirsuthum*); these severally turn the edge of the scythe, and lie an unused lumber where they fall. The spade or the plough may thus eradicate the encumberers of the ground for ages past : perhaps a second or cross ploughing may be practised, before the late autumnal rain prevents that operation, and subjecting to frosts the roots and seeds of their various tribes. In fine, the turned up soil may be harrowed once, twice, or thrice, by the middle of April. But before I proceed, I shall take the liberty of mentioning my own experiment with two portions of low meadow, that were not only flowed over by the winter floods, but on which water rested during the spring months. A graduated channel to an adjoining run of water, drained much of the water away ; but from the long habit of its resting, most of those weeds mentioned by Mr. Davis, domineered over the scanty grass that grew. I opened ditches, ploughed
them

them as I have proposed, and laid potatoes in sets on the harrowed surface, with a small portion of rotten dung on each set. I opened pathways by shovelling the mould, till these prominences appeared like ant-hills. Indeed, both parties were considerably productive; their product has been in request, as preferable in flavour to the strain from whence they were raised. I may be excused repeating, that the greatest part of these grew where only *beggary* (our provincial term for such weeds as have been adverted to) grew before.

" There is no doubt with me, but that the Oxfordshire meadows may be thus reclaimed. Potatoes are now in much estimation for man and beast; but it has not appeared to me that they are frequent in that county, where its inhabitants may find occasionally the best succedaneum for bread; and if they grow a superfluity, it is well known that swine may be aided, if not fattened, with them, and the bullock equally promoted. Besides, the new Canal, now led through two-thirds of the county, renders a freightage easy to the metropolis, where they are in certain demand. To finish this long article: after the requisite summer-hoeings for the potatoes, clean crops of wheat, oats, and even barley, may be obtained ; or if there is an object that they should revert again to pasture, selected seeds from the best native grasses and trefoils will prosperously grow on the purified soil, more abundant, as more unblended with any unwanted production,"— *MS. Annot. Mr. Wagstaffe, of Banbury.*

SECT. II.—PASTURES.

" The enclosed pasture or meadow land, is chiefly confined to the middle part of the county, near to Oxford, where there is a pretty large tract of deep rich soil; from the pasturage of which, besides the quantity of butter made, which is considerable, a great number of calves are suckled, and the veal sent to the London market. On various parts of this district there are some oxen, cows, and sheep fatted. As much of this tract of land lies wet, a very great improvement is experienced by under-draining both pasture and arable, which is done in various methods; though none is so generally approved, or so durable, as the drains that are made with stone, where it is to be procured. The process of under-draining being expensive, it is not so universally practised as it might be, especially where tenants have no leases; but I have often known landlords contribute to this expense. The Essex mode of under-draining, which has been introduced in some places by gentlemen on their respective farms, is found to answer best. Many of the pasture grounds are full of ant-hills, and the herbage growing thereon is coarse, and refused by the cattle, who will be much reduced for feed before they will touch it, even when the young grass shoots up amongst the old, which is dry and withered * "

The best feeding land is found on the banks of the three rivers, the Thame, the Isis, and the Cherwell; but the lowest meadows are subject to floods, which

---

* Original Report.

.　　　sometimes

sometimes causes much damage, by coming too late in the spring.

The grass lands at Waterperry, under cows, are 40*s.* to 50*s.* rent.

At Waterstock, enclosed pasturage for cows chiefly prevails: in general, the application is to the dairy. The grass lands are good, and lett for miles together, the meadows at 50*s.* and the pasture grounds at 40*s.* They mow two tons an acre the first growth, and one ton the second; but these second crops reckoned bad for the land.

SECT. III.——LAYING DOWN TO GRASS.

Mr. Blake, at Hampton, has laid down a field with grass-seeds, sown alone in August: the success very good.

Captain Lechmere, of His Majesty's ship Dreadnought, laid down a field to grass at Middle Aston; and being advised to mow it the first, second, and third year, followed the advice, and nearly spoilt it. It was fed for one or two years more, and pared and burnt for turnips; after which, barley; and with that crop, crested dogstail, and some other clean separate seeds. It has been fed this (the first) year, and any person would suppose it had been down seven years, so full, thick, and luxuriant is the crop. The business is done for duration, and well done.

On the enclosure of the parish, Mr. Edmonds pared and burnt an allotment of the common, and took two crops of turnips in succession: the soil, a moory surface on a gravel bottom. Fed off with sheep, and

sowed

sowed with seeds in May, without any corn, rye-grass, hops, or honeysuckle. Very good keep in autumn : afterwards, abundance. There are spots where the bottom is malmy, and not gravel—not porous for the water to get off : on these the grasses do not flourish so well, nor come so early.

Mr. Warrener, at Bloxham, in 1806, laid down two acres and a half with meadow fescue ; and with the seed of it, fourteen more this year (1807) : Dutch clover with it.

Mr. Foster, at Bignal, this year (1807) laid down a large field to grass, by fallowing till July, and then sowing white clover and hay-seeds. I did not see an omission in the part of his farm that I walked over, but the neglect of not running over this field to cut off a few weeds, which were left to rise too high : it has a full plant of grass.

———

As the plants of a new lay come up weak the first year, does not this weakness of the plant proceed from the small quantity sown ? Is it advisable to sow corn amongst it ? Will corn, by being sown amongst the grass-seeds, occasion them to come up stronger; and will it not rather draw a part of that nourishment which ought to go to the support of the grass ? If weeds grow up, will it not be best, if they are such as charlock, to draw them up by the hand; and if docks and thistles, by means of an instrument; and can these operations be performed so well with a corn crop ? I incline to the opinion of laying down grass-seeds without a crop of corn, according to the plan of Mr. Majendie, related in the Annals of Agriculture.—*Dr. Sibthorpe.*

In dry stonebrash ground, such as we observe about Burford, might not the *Cynosurus cæruleus*, or blue-crested dogstail-grass, be introduced with advantage ? This flowers the earliest of all our British grasses, and, from its appearance in the garden, promises to furnish a good crop.—*Dr. Sibthorpe.*

The number of British plants growing in Oxfordshire does not amount to more than 1200 ; of these, a very considerable part has been classed by botanists under the title of *Cryptogamia.* Some of these are the object of agricultural inquiry. The number of British grasses is not very large,
and

and the knowledge of considerably less than a hundred plants, would enable the agriculturist to make experiments with botanical accuracy. Thus, for want of this botanical accuracy, we are often told, in our best writers on agriculture, that burnet grows as well on wet as dry land. This confusion has arisen from the same English name having been given to two different plants. The *Poterium sanguisorba* grows on dry calcareous soils, on which the sainfoin delights; the *Sanguisorba officinalis* grows in moist meadows, or such as have been occasionally flooded.—*Dr. Sibthorpe.*

A new species of *rye*-grass—would not this word be more accurately written *ray*-grass; as the name is taken from the French *ivrai*, from the intoxicating quality the seeds of some sorts of this kind of grass possess? The term rye-grass seems to denote its being a species of *Secale*, or rye, to which it has no affinity.—*Dr. Sibthorpe.*

CHAP.

CHAP. IX.

GARDENS AND ORCHARDS.

———◆———

SIR C. Willoughby has an orchard which occupied ten acres, and at present nine, planted about 70 years ago. He has constantly made cider, except for one year, when he sold the apples for 100*l.* In 1806 he made 29 hogsheads. Six years ago, one tree gave three hogsheads, and another was as good. The apple is the broad-nosed pippin; the produce, on an average, is above 20 hogsheads. The advantage has been such, as to induce him to plant five acres more.

Mr. Knapp, M. P. for Abington, has an acre and a half, the apples of which sold this year (1807) for 35*l.*; another, at Kidlington, for 35 guineas. Last year the price was 30*l.*; but the year before, no crop. In one of these orchards, a single walnut-tree is reckoned at 5*l.* 5*s.* of the money.

The Earl of Shrewsbury's conservatory is very great, well placed, and sheltered from every thing but the sun: 248 feet long; the central division 80 feet long, 30 wide, and 15 high. It is highly productive: 22 peach and nectarine trees have yielded about 60 dozen on each tree, or above 15,800, and peaches of 14 oz. weight: the vines, 6800 bunches.

CHAP. X.

WOODS AND PLANTATIONS.

THE produce of the woods at Stokenchurch is 40 per cent. dearer than it was 20 years ago, arising, in great measure, from the number that have been grubbed up. The motive has been, the conversion of the stock growing on the land into money : the interest of the price, and the rent of the land in arable, together pay far better than the wood.

At Wormsley, the large beech-trees measure as timber, and the smaller go as poles of half a load, or 25 feet : these may be 20 or 25 years old. Twenty years ago, the price was 8*s*. 6*d*., now 24*s*. the half load. Their profit now is considerable; but if there is a fair stock of trees, so is the temptation to grub : for a common offer is from 30*l*. to 35*l*. per acre, and to leave the land grubbed for the plough ; thus the produce is sold at 30 or 35 years' purchase, and the land kept. If the money is invested at five per cent. the profit, in such cases of grubbing, is decided at the first blush. But some woods are on such steep declivities, that the land is good for little in any other application : the account, however, stands good, as cutting *all* does not necessarily imply grubbing: it may continue, wood demanding patience.

Mr. Fane pays great attention to the trimming of his trees, to lead and keep them as erect as possible, that
a spread-

a spreading growth may not drip on, and check the younger succession which are saved from seedlings, if possible. The best time of doing it is that of thinning: the value of the faggots more than pays the expense—6*s.* per 100, and they sell for 12*s.* He has invented a very useful tool for trimming small spray—a bill-hook on a long pole, the end of which is a blade, and cuts; this takes them off smooth, without any tearing.

Sir C. Willoughby has 300 acres of beech woods on the Chiltern hills, which are cut at 40 years growth: they are thinned once in seven years. The price at which the wood is sold has risen, in 20 years, from 9*s.* 6*d.* to 22*s.* When the woods are bought for grubbing, the purchasers, though they contract to do it, do not in fact half effect it, and therefore it must be done after them. The price is not more than 21*s.* per load, but varying by nearness to the river. These woods are not titheable; but this depends, in a measure, on the old stools being grubbed up: for underwood (legally, that which shoots from a stub) is titheable, whatever the age may be.

The Bishop of Durham, at Mungwell, possesses beech woods, which are managed in the thinning system at 20 years growth, and pay him about as well as letting the contiguous arable land. No woods of any extent have been grubbed up in that vicinity. Although some Oxfordshire decisions have exempted beech woods from poor-rates, his Lordship continues to pay them, disapproving of such exemption.

The present Duke of Marlborough planted the great belt at Blenheim, the extent of which is 13 miles: amongst all the plantations there, I remarked a great want of thinning.

Mr. Pratt informed me, that the Duke of Marlborough's

rough's woods are in general cut at from twelve to fourteen years growth : sold at 40 *braid* the acre, or four poles square—sold all standing. A price fixed on every lot, and they draw lots for it ; they do not see or know the price, only the number of the lot, and the number of the braids. From 1s. 6d. to 5s. a braid, the buyer cutting it, and doing what they please to clear it by Midsummer, and pay for it then from 3l. to 10l. an acre ; then draw the timber. To leave samplers tellows, twelve to the acre ; allow 1d. each for all left. They must grow from acorns, on account of making the best timber ; and every year plant in November, by the side of bushes of last year's growth ; and they are drawn up, and do best. Did this from observing in the forest, that a bush often nurses up an oak. The forest does not pay more than 6s. per acre : he is of opinion, that it would answer to grub up many ; the expense, 1s. a pole, or 8l. per acre. He has had it done, and cultivated, first crop, oats, seven or eight quarters ; then beans, but blighted, and much straw ; and two quarters and a half of barley, and very good ; then seeds, good and great one year ; then wheat, and good. The roots pay about one half. Mr. Brown, of Kidlington, did it also, and it answered well on 20 acres.

The late and present Mr. Stratton planted at Tew, on a scale and plan that decorates that fine estate considerably ; the amount is now 63 acres, and no year passes without additions. There are many thousand trees of 15 years growth, from 20 to 30 feet high.

Mr. Coburn, of Witney, has many willows in his low meadows ; and he is very careful in pruning them the second or third year of the growth of their heads, clearing out all the under-shoots of straggling spray, in order that the tops may consist solely of clean poles,

and

and these grow larger by means of this management—an object of no small consequence, as the poles sell, at nine years growth, from 1*s*. 6*d*. to 2*s*. a pair, for enclosure rails.

" The woodlands in this county may be divided into three sorts, viz. *groves*, or *spring woods*, consisting of trees only; *woods*, consisting of timber trees and underwood ; and *coppices*, consisting of underwood only. Of the first of these descriptions may be considered,

" The *beech woods* of Oxfordshire, which are confined to the Chiltern country, and consist of trees growing on their own stems, produced by the falling of the beech-mast ; as very little is permitted to grow on the old stools, which are generally grubbed up. They are drawn occasionally, being never felled all at once, except for the purpose of converting the land into tillage, which has been much in practice of late years. The beech wood thus drawn is either sold in long lengths, called poles, or cut short in billet lengths, and sold for fuel. It requires some judgment to thin these woods, so that the present stock may not hang too much over the young seedlings; at the same time that, in a south aspect, an injury may take place, by exposing the soil too much to the sun : for it is to be observed, that the north side of a hill will produce a better growth of beech than the south side, the very reverse of which is the case with regard to corn. The succession of young trees in beech woods is much injured, by admitting sheep or other cattle into them ; and though it is said by some, that sheep do no damage in winter, when the leaf is off, and find considerable feed from the grass and other plants abundant in woods, yet it is the opinion of others, that the wool

which

which is left hanging on the young stock, is prejudicial to its growth, supposing what is doubtful, that
the sheep do not crop them. Some improvement might
be made by keeping better fences, particularly against
commons, where a wide ditch is an essential part of
the mound; and by transplanting the young beech
from those parts of a wood where they are too thick
(so that the strongest would destroy the weakest), to
those places where they do not stand sufficiently thick,
there being spots of both these descriptions to be found
in most woods.

" There are some oak and ash trees in these woods,
dispersed among the beech, which have sprung up in
such places where the seeds have dropped, or been
carried by birds, or other means. These seldom grow
to any great bulk, though sometimes to great lengths;
but they are not very numerous.

" Of the second kind are the *woods* in the vicinity
of Stanton St. John, called the *Quarters*. The soil
here being a strong clay, is well adapted to the
growth of oak. There are many spots of woodland
of this description dispersed about various parts of the
country. Coppices do not abound in this county : indeed, there are very few of any extent, except those
called coppices in the forest of Whichwood; though
these, having trees in them, are more properly woods.
I would beg leave, on the ground of some experience,
to recommend to gentlemen the cultivation of coppices
on their respective estates. Many spots might be
found that would turn out to great advantage, by
being applied to this purpose; and the rather, since
there can be no doubt but, in proportion as improvements take place in husbandry, a greater stock of
sheep will be kept, and of course, there will be a

OXFORD.]   greater

greater demand for hurdles used in folding them, **and** for other fences. It is worthy of notice, that on **the** moist part of the land, ash and willow may be **culti-** vated to the greatest profit *."

---

* Original Report.

CHAP. XI.

WASTES.

EXCEPT the two large tracts of Whichwood-forest and Ottmoor, the waste land in Oxfordshire is not very considerable.

Ottmoor.—Mr. Weyland, of Wood Eaton, had the goodness to shew me this large tract of waste, in a manner which I could not have done but through his attention, who requested a farmer (Mr. Jones) who turns much stock on it, to conduct us in a little tour over the whole of it. Mr. Jones conceives it to equal the amount of four square miles, or 2560 acres: it is a very singular tract, nearly upon a dead level, surrounded on every side by high lands. I searched and inquired for bog and peat, but have reason to believe that there is no such thing upon it, or if there be, in a very small proportion. The soil is generally a good loam, which would form most valuable farms, and be applicable either to tillage or pasturage, were it drained; but more adapted to the latter than the former. To the eye it is level; but there is inequality of surface enough to make the waters draw down to one part of it more than to the rest: so that part of it is very wet in October, after a long series of fine weather, and the rest quite a dry sound turf. The river Ray runs across it; and apparently, there would not be any difficulty in draining it. — Croke, LL. D. of Studley, who has considerable rights upon it, before

he

he went to India, ordered some levels to be taken, in order to examine the best means of draining it; whether by clearing the river Cherwell of obstructions, or by a new cut through Wood Eaton, or by a cut to carry the water to the Thames: but I could not hear what the result of these inquiries was. The scheme, in general, was hotly opposed; and Lord Abingdon, who is paramount over the seven towns, so far as this moor is concerned, being adverse to the plan, nothing came of it. I made various inquiries into the present value of it by rights of commonage; but could ascertain no more than the general fact, of its being to a very beggarly amount. Mr. Jones is in the constant practice of summering some cattle, that are bought in for this purpose; and in a rough way estimated, that he made about 20*s.* a head, on an average; but subject to the hazard of losing a beast, which happened now and then by distempers. Sheep are liable to the rot; and Mr. Jones himself lost one whole flock of 17 score upon it, saving only three from that number. The number of horses, beasts, and geese which I saw upon it, was very inconsiderable. Upon the whole, the present produce must be quite contemptible, when compared with the benefit which would result from enclosing it. And I cannot but remark, that such a tract of waste land in summer, and covered the winter through with water, to remain in such a state, within five miles of Oxford and the Thames, in a kingdom that regularly imports to the amount of a million sterling in corn, and is almost periodically visited with apprehensions of want—is a scandal to the national policy.

The following towns have the right of commonage upon the moor: Charlton, Oddington, Noke, Beckley, Horton, Fencott, Mercott. No sheep have been turned

turned on it this summer (1807), fearing the rot; but cattle never did so well on it. The *flits* upon the moor, are the holes whence peat has been dug : *pills*, are hills of quaking bog.

If drained and enclosed, it is said that no difficulty would occur in letting it at 30*s*. per acre, and some assert even 40*s*. The subject being of much importance I shall add Mr. Davis's remark.

" There are in most of the unenclosed parishes, either small or larger tracts of wastes, or down land, which are appropriated chiefly to the feed of sheep : the range of Chiltern hills which cross the southward end of the county, are of this description, being in many places too steep to plough. In the more northern part of the county, there are considerable tracts of down land belonging to most villages, which are often overrun with ant-hills and coarse herbage, being of little value, and chiefly appropriated to the pasturage of young cattle, or sometimes, where they are good enough for that purpose, and sufficiently extensive, of oxen for the use of the plough.

" The most considerable, and at the same time most valuable, tract of waste in this county, is the common of Ottmoor, situated near Islip : which contains, as near as can be ascertained, about 4000 acres, and is commonable to eight adjoining townships. This whole tract of land lies so extremely flat, that the water, in wet seasons, stands on it a long time together, and, of course, renders it very unwholesome to the cattle, as well as the neighbourhood : the sheep are thereby subject to the rot, and the larger cattle to a disorder called the *moor evil*. The abuses here (as is the case of most commons where many parishes are concerned) are very great, there being no regular stint, but each neighbouring

bouring householder turns out upon the moor what number he pleases. There are large flocks of geese likewise kept on this common, by which several people gain a livelihood.

" It was in contemplation a few years ago, to drain and enclose Ottmoor ; and it is a great pity such a valuable tract should not be improved to the utmost, for the advantage at once of the occupier, the proprietor, and the public.

" It is not easy to ascertain the quantity of the other waste or down land in this county ; but it must be great, as there remain at this time upwards of 100 unenclosed parishes, or hamlets, to which there are wastes belonging in greater or lesser quantities, although in most of them the commonable rights are stinted*."

" The expediency, I may add, the necessity, of enclosing, is happily adverted to by this gentleman on this large tract of waste. I knew a farmer, whose sheep were sound and healthy for a succession of years, but who, in a dry season, permitted their roaming to a watry waste which was of small compass : what plants they ate were not ascertained ; there grew numerous plants whose spontaneity is frequent in such places : but shortly after they were removed, they discovered the symptoms of decay, viz. the rot. The farmer anticipated the consequence, and sold them at a low rate, for present slaughter : since when, this space has fallen to my lot. I have extirpated the aboriginal plants, carried on some new earth, and varied fruitful crops have succeeded.

" I may just remark, that the draining of such places may be as essential to the health of man as

---

* Original Report.

beasts.

beasts. I deem it may be perceived, that **intermittents,** or the ague, hath nearly lost its hold on the human constitution; at least, it is acknowledged so in this vicinity (Bawburgh); and I am much disposed to believe, that a universal drainage through any district, would tend to its annihilation there. Passing or living waters, I conceive, contribute to health; but this disease, and perhaps some others, originate from stagnant waters, and more peculiarly from the disgustful effluvia arising from their weedy beds in sunny weather, after evaporation: which baneful effluvia even the muddy tracks of rivers send up, when, in very dry weather, they are forsaken by the current. Hence there is a **two-**fold inducement to remove the means that vitiate the air, when the subject removed induces a fertility in the fields of cultivation."—*MS. Annot. Mr. Wagstaffe, of Banbury.*

There is a very large common at Kidlington, which feeds 300 cows from the 16th of May to Michaelmas, all by the farmers, according to their yardlands and their cottages, but stinted by agreement to three-fourths of the right; that is, to turning on three cows instead of four, for four common rights, each right being for a cow: and this cow-right letts for two guineas, liable to tithe, and have sold for 60*l.* Campsfield is a sheep-common in the same parish with adjistment shepherds. A man hires common rights, and he provides the sheep, and pens arable land, from the 30th of March till November: then they go to the cow-common and the common meadows, buying hay for the depth of winter. Both sheep and shepherds miserably poor: the breed, a mixture of long-woolled sheep, Leicester, Cotswold, and Berks; many have black faces. The cow common becomes a horse-common from after harvest to the

the 5th November ; then the sheep and cows go into the common meadows and the stubbles.

Mr. Taunton, Clerk of the Peace for the County, has been for some years in possession of between 300 and 400 acres of allotment of an enclosed waste at Ensham, which he has been for some years gradually bringing into cultivation. There are two sorts of soil—one, a yellow loam of various depths, strongly tending to clay, except on the surface, and burns into brick; and a lighter and more porous and friable loam, tending to peat, but wet, and requires hollow-draining, which has been largely executed by Mr. Taunton : this peaty loam is by far the best land. The spontaneous growth of the whole tract was furze, high and luxuriant, and mixed with fern. The best land by far is the peaty, and the yellow loam the worst; nor are many soils to be found of a worse complexion than yellow loams, when they want any draining.

The first operation necessary was, to cut the furze ; Mr. Taunton gave it to all who would clear and *grub*. This was an error at the commencement of the works ; the furze was of considerable value, and no slight profit was made by the farmers and others who thus contracted for the work : they sold the furze, but grubbed so imperfectly, that much of it was to be done again. Nor was this the only evil ; it was afterwards found, that lime was a powerful manure on these wastes, and Mr. Taunton erected four kilns to burn it ; one for lime and brick at the same time (one of these for furze and three for coal) ; but most of the furze was gone which should thus have been supplied. These errors were for want of experience, which must be *bought* by every man who does not avail himself of the knowledge already gained, and registered in books.

Furze

Furze is cut at 3*s.* to 3*s.* 6*d.* for 120 faggots; and they sell from 13*s.* to 15*s.* being taken away by the purchaser : the roots are grubbed at 4*s.* per stack ; the bailiff knew not the proportions per acre. The land was then pared and burnt, and the ashes spread ; the expense, 30*s.* per acre; and then the land was ploughed, and oats sown. The course after this was,

1. Turnips,
2. Oats,
3. Ray-grass and red clover,
4. Ditto,
5. Oats.

The lime spread, 25 to 30 quarters per acre, for the turnips, and also for any other of the crops previous to ploughing.

But potatoes have made a variation : the land being pared and burnt, labourers, little farmers, &c. gave 3*l.* per acre, for liberty to take a crop of that root; and some Mr. Taunton planted on his own account, the crops varying, according to the humidity of the year, from 30 to 100 sacks per acre. After potatoes, oats, and the crops great : this is the crop which suits the land best ; wheat and barley have been tried, but failed; the land is left by furze too loose and hollow for wheat. Oats succeed extremely well after liming, and produce from five to eight quarters per acre. Sows four bushels per acre ; but of the potatoe-oat, three bushels and three quarters. Drags and harrows in according to the soil, and has ploughed in some. But after a round of seven or eight years, it is expected that wheat will do well on the layers of two years. As the quantity of ray-grass sown is considerable, and left two years, Mr. Taunton may be disappointed in this expectation. Oats, on such a layer, is safer husbandry, dibbled in ;

or

or pease, which like to follow ray, and then the wheat; taking care not to plough too often in any case, when wheat is intended. He mows the seeds, and gets very fine crops of hay; this year, two tons per acre the first mowing.

The effect of the lime is very great, and seen to an inch, and kills fern and all other weeds; it is much more effective than dung, and lasts twice as long. He burns off 36 quarters at a time in coal-kilns. For brick and lime together, the arch and basis to filling half the kiln is built of limestone, and the bricks upon it at top.

Draining Mr. Taunton has practised here largely. Digs from 12 to 16 inches deep, and filled up at first with stones tumbled in; but this being ill done, and not proving effective, he has since laid them carefully to form an open channel: the price, 8*d.* a pole of six yards.

The sheep-fold is another mode of improvement applied to this farm. He has at present of breeding ewes, lambs, shear-hogs, and wethers, 190: has preferred the Berkshire breed, but is now adopting a change, by mixing with the Cotswold. The Berks tod, six or seven; the others, three or four. Has sold Berks shear-hogs, in store order, at 43*s.* a head.

I cannot but remark, that the number of sheep kept is inferior to what a new improvement demands: in all such improvements, sheep form one of the principal tools with which an improver should work.

Stock.—Nine horses; 18 cow kine in winter, none in summer: no swine. Rabbits, a great nuisance on the farm, and will be till all the furze is destroyed; But a neighbour (Colonel Parker) killed 5000 couple in one year, which ought to benefit the whole vicinity.

Crops,

Crops, 1807.

Turnips on stubble	....	17 acres.
Oats	... .	86
Turnips	....	28
Fallow	....	15
Grasses	....	48 acres mown.
Ditto	....	22 ditto fed.
Potatoes	....	18
Osiers	....	12

$$\overline{229}$$

The rest furze.

Mr. Taunton is building a new house here, where he probably intends to reside, as it is too good for a farmhouse.

" I think it would be a prosperous circumstance to this county, as well as every other of the same description with extended downs, or sheep-walks, to make a temporary experiment with a course of crops, by hurdling in (if there is a doubt of its permanent success), here and there, ten or twelve acres of down ; and, with a single ploughing (such as is frequently practised on old lays in Norfolk), turn up the turf about eight inches over, and thereon dibble in parallel rows about four inches asunder, and the holes about one inch and a half, for wheat ; or a single row for pease, the holes about two inches and a half apart. When the crop is carried, the hurdles may be thrown open, and the sheep renew their range over the shorn stubbles ; in which state it may continue till mid-winter, or, indeed, till the fourth or fifth month of the ensuing year ; when, there is not a doubt with me, but if an immediate tillage

lage takes place for turnips, a larger tonnage of pro-
vender will be in store for the winter feeding, than hay
which might be made, or grass that might be cropped,
would have amounted to.   When the turnips are ate
off, a barley or oat seeding becomes requisite, with
which may be sown, if it be judged proper that this
cultivated piece should revert back to sheep pasture, to
sow it with seeds selected from the grasses predominant
before breaking up ; or, if a year or two of hay is the
object, then the trifoliums, &c.   Under these circumstances I do not hesitate to pronounce, that the land
will be worth double rental for four or five successive
years ; and should cultivation be suspended, the change
effected will eventually produce an increase of herbage
through many subsequent years."—*MS. Annot. Mr.
Wagstaffe, of Banbury.*

WHICHWOOD-FOREST.

Abstract.

		Statute Measure.	
	A.	R.	P.
King's coppices	1649	2	10
Baron's ditto (Duke of Marlborough)	1041	3	17
—————— (Mr. Fettyplace, &c.)	346	0	33
Keeper's lodges and lawns	134	0	23
The open forest	2421	1	15
	5593	0	18
The chase woods	487	3	4
Blandford-park	639	2	17
	6720	1	39

The rights of common upon the forest are for horses
and

and sheep only—no cattle or hogs; but the number of both, by trespass, are very great.

Soil.—The soil of the forest, over which I rode above 16 miles, in order to examine every part of it, I found to be either a reddish good loam, or the common stonebrash of the extensive district in which it is situated. At one spot, where some ponds were choked up with weeds and mud, there is found peat earth, which Mrs. Stratton has used very successfully for her exotic plants demanding that earth, and is a hint of the origin of bogs. The whole forest, were it enclosed, would lett at least as well as the surrounding country, which (except certain properties) runs at 20*s.* per acre.

Copses.—There are 34 copses in the forest; 18 of which belong to the King, 12 to the Duke of Marlborough, and four to certain individuals; these may average about 100 acres each. The King's are cut at 18 years growth, and the Duke's at 21. There is a right, and consequently the practice, of fencing these off when cut, by a hedge and ditch, for keeping out all commonable cattle and sheep for seven years; after which, all must enter: the deer are never excluded. These copses return, on an average, about 6*l.* per acre, clear: the cutting costs 40*s.* and the faggoting 2*s.* 6*d.* per 100; but these expenses are paid by the purchasers. Eighteen years producing 6*l.* is in the ratio of 6*s.* 8*d.* per acre per annum. The open forest produces nothing, but a certain quantity of brush-fuel, and browse for the deer.

Timber.—I did not see one very fine tree of navy oak in a ride of 16 or 17 miles; but a considerable
number

number of thriving ones, which appear to be of 60 or 70 years growth, and which promise in 100 years more to be valuable ; but when these trees are compared with the space of land in which they are found, they cease to be an object of any consideration. Next to oak, ash seems to abound, and then beech, with some elms. There is not the smallest reason to judge from the appearance, that, in its present state, this forest will ever be productive of navy timber in the least degree answerable to its extent.

Enclosing.—In riding over the forest, I found many very beautiful scenes, particularly where the *nut* fair is held, a glen by Mr. Dacre's lodge, and others approaching Blandford-park, &c. : there are vales, also, of the finest turf. Several of these scenes want nothing but water, to form most pleasing and finished landscapes. The present ranger, Lord Francis Spencer, has made several roads, by way of ridings, through the forest ; and no person can doubt, but that to the residents who live on or near it, this fine wild tract of country affords many agreeable circumstances which may operate, in a measure, to prevent an enclosure which ought, for a thousand reasons, to take place as soon as possible. No reasonable person can propose or desire, that a measure should be effected to deprive any one of any sort of property ; the object is merely to make a large tract of good land productive to the public : to whom it shall be apportioned is not of the smallest consequence ; the cultivation is greatly to be wished, and this would take place on its being made private property, in whatever hands it might be placed. The person whose feelings would be most apprehended (the ranger), is fortunately of much too liberal a mind

to

to oppose any measure really and substantially for the general benefit; and there are no circumstances (the mere pleasure of wandering alone excepted) which may not be fully compensated, by the solid and valuable consideration of allotments.

Nor is it in the view of productiveness alone, that such an enclosure is to be wished: the morals of the whole surrounding country demand it imperiously. The vicinity is filled with poachers, deer-stealers, thieves, and pilferers of every kind: offences of almost every description abound so much, that the offenders are a terror to all quiet and well-disposed persons; and Oxford gaol would be uninhabited, were it not for this fertile source of crimes. This is a consideration that will surely have its weight with every man who sees the evil, and must consequently wish for the only remedy the case admits.

A fact which bears also on this point, and flows from it, is a circumstance mentioned by Mr. Pratt, of Fowler; that poor-rates in the parishes which surround the forest, and have or usurp a right of commonage there, are higher than in others under similar circumstances, except in that of being cut off from the forest.

Open Fields.—In the open fields in the district around Baldon, there are no division baulks, which in so many counties are sources of weeds and depredation: if encroachments are made, the matter is settled by a jury. It is an acknowledged and well-known fact, that men have ploughed their land in the night, for the express purpose of stealing a furrow from their neighbour; and at all times it is a constant practice in some, to plough from each other. Such a practice is

some

some apology for tacking this article to the Chapter of *Wastes*.

I add with pleasure the following valuable communication from W. Lowndes, Esq.

" The late enclosure of Brightwell, in 1800, having added 225 acres to the cultivation of the country, may not be unworthy of remark ; though little can be gathered that may lead to the grand secret of cultivating such wastes as possess no great depth of soil. Finding it necessary to build an entire homestall, the winter approached before I began to cultivate ; and having received from the neighbouring farmers a proposition (which, for their honour, ought not to pass unnoticed), gratuitously to lend me their teams at convenient times during the winter, I immediately ordered the furze to be grubbed, which I contracted for at from 8*s.* to 12*s.* an acre, and completed about 100 acres during the autumn, which, for want of sale, was burnt upon the ground. Having provided four ploughs, my bailiff, in their turn, gave notice to my neighbours, who cheerfully with their teams attended, and ploughed the whole 100 acres, in due course sown with oats : the remainder of the common I pared and burnt, and got the whole into cultivation by the end of the third year. Neither process exempted me from the trouble or concern of eradicating the fern by the hoe, which appeared in as luxuriant a state after ploughing as before ; and my experience since has convinced me, that nothing can eradicate it but continued cultivation, followed by the hoe.

" On the part not burnt, my process has been generally, oats, turnips, wheat, and clover or tares : on the part burnt, the process has been, wheat, turnips, oats, clover or tares ; the latter I prefer. The wheat

was

was clean and full grained, and equal to any expectation I could have from that soil, and sold for seed ; the turnips luxuriant ; the oats of the potatoe kind, a full crop ; the clover bad throughout, and the tares moderate. I had enemies in succession to defeat : 1st, the fern ; 2d, the seedling furze ; and 3d, the grasses. The fern, each crop, grew weaker and weaker, and ceases to be an object : the furze-seeds seem hitherto to multiply on each ploughing, and in all crops of continuance, the grasses likewise. I have therefore come to a determination to sow no more clover till my enemies disappear ; and I am prepared also to cease the wheat crop, if I find the difficulties continue : nothing but crops of short duration, and immediate turning up the soil, can destroy either of the latter causes. The means which, under the process stated, I have taken to destroy the grasses, is by breast-ploughing them, and forming them into composts with quick-lime, which I burn on the farm. In about three months the heaps are fit for use, the whole of the vegetable being reduced into a black powder, and the lime completely slacked, so that the compost may be spread on in any quantity thought proper. The improvement is two-fold—by eradication, and fertilization, and the benefit too obvious to dwell upon.

" Before the enclosure, the common had sufficient plots or intervals of sward, to maintain during summer the sheep of the parish, receiving some additional food at night in their respective folds. On all parts where the sheep depastured freely, and in proportion to the depth of the soil, I find less inconvenience and better crops ; but there are patches—some where the rabbits (the former occupiers of the soil) had burrowed, and some where the furze grew so thick as to destroy the sward, and exclude the sheep—on which spots all the crops fail. On one of them, which appeared to me most inveterate, after paring and burning, I had twice folded, covered with lime to the extent of 80 quarters per acre, and well dunged with stable-dung, yet the effect was the same; a pretty remarkable one, of converting every grain into grass, and leaving a bed of the *Agrostis* irradicable. Yet I do not despair of conquering it : on the others, my hopes are more immediate and sanguine.

" The soil is uniformly of the same quality, light and friable, but of different depths; the subsoil is clay, mixed with flints, and occasionally a layer of gravel. After digging about 12 feet, the soil changes into a chalky marl, varying in quality, but all of it useful in cultivation, and of which I have begun to avail myself, and expect great benefit.

" From the experience I have had, I would begin on such a farm with the following process : after grubbing, paring and burning, 1. turnips, 2. winter vetches, 3. rye, 4. turnips; all of which being fed off with sheep, and occupying the period of three years, might be succeeded by oats, with the usual crops in succession."

CHAP. XII.

IMPROVEMENTS.

SECT. I.——DRAINING.

THE scattered tracts of land in this county that demand draining, when compared with the districts which do not require this improvement, are not considerable. The method of hollow-draining, when necessary, is pretty well understood all over the county; and as the drains are usually filled with stone, when once made they are made for ever.

Mr. Latham, of Clifton, has made very laudable exertions in under-draining, to the amount of some hundreds of pounds. As the evil results from springs, the depth, of course, varies according to the necessity of the case: the common depth is three feet; and they are filled with furze, blackthorn, or the green spray of elm, which last is best of all, blackthorn good, and furze is durable. In all cases covered with wheat-straw. Stones would be better than either of these; but the carriage would render the expense too great. The price varies from 3*d.* to 6*d.* per rod, and all expenses included, amount to about 1*s.* per rod. the drains are cut from 10 to 15 yards asunder, across the slope of the field; and main ones disposed, where proper, for the others to empty into.

Mr Turner, of Burford, has expended 200*l.* in hollow-draining : cuts them two feet deep, and upon
stonebrash

stonebrash land, much deeper, to take off springs; they are filled with flat stones, so as to leave a hollow for the current. At 24 inches deep, the expense from 1*d.* to 1½*d.* per yard.

Mr. Foster, at Bignal, whose farm speaks him to be one of the best cultivators in the county, has made very great exertions in draining. His soil is a stonebrash, upon a blue clay and white marl, and under this a white limestone, and springs abound in it enough to render hollow-drains necessary. I found a gang of men digging them four feet deep, and mostly two feet broad, with a cut at bottom one foot wide, in which broad stones are so disposed, as to afford a free open channel for the water: at bottom, they have to quarry in the limestone, and to blast it with gunpowder. They work by the day. I asked them what they would contract for such work at; they replied, 7*s.* to 8*s.* per rod of six yards, and could not do it for less. This is a great exertion, and Mr. Foster has done a great deal of it. One circumstance I could not but take particular notice of, as I had never seen it before: the ditches seeming to be very shallow, I found, on examining them, that a hollow-drain was made at the bottom of them, and then filled to the depth I saw. The reason is, that if they are left open to a sufficient depth to take the water away, the sides shatter down, so as soon to stop the drains. Admirable management! in praise of which too much cannot be said. Some of his drains, after a long course of dry weather, were running a perfect stream. I wonder where a tenant is to be found, who would make such exertions while tenant at will, or on a short lease!

SECT. II.—PARING AND BURNING.

This husbandry is scarcely ever practised on the hills of Wormsley, or in any part of the southern division of the county : this may possibly result from some ill-made trials. Thus, at Stokenchurch, 25 acres of clovers that had been down only six or seven years, were pared and burnt ; but the farmer knew so little of the husbandry, that he rafter-ploughed it for turnips, and on that half tillage, on a lay of insufficient duration, sowed his turnips : they failed.

Mr. Lowndes, on the new enclosure at Brightwell, pared and burnt 38 acres of the common allotted to him, in 1802, which was sown with wheat, and proved very good ; succeeded by turnips, which proved good also. Oats followed, a fair crop ; then very fine clover on some parts ; on others, middling : it now lies for wheat.

Sir C. Willoughby pared and burnt 20 acres of imperfect boggy bottom, and planted potatoes ; the crop very great. A great crop of oats followed, and eight acres of it with them was laid down to grass, which proved good ; but got coarse, and must be ploughed again. The other twelve acres were burnt for turnips, which were good, and followed by a great crop of oats ; after these, turnips were taken again, which were good for the year.

In the allotment of the Rev. Mr. Dupuis, on the enclosure of Wendlebury, he had part of an old common which he pared and burnt, and spread, at 28s. an acre, in the spring : ploughed and cross-ploughed, and sowing turnips, they failed. He then ploughed

again,

again, and got the finest turnips that were to be seen in the country. These were fed on the land by sheep, and the oats which followed were eight quarters an acre.

In 1807 he pared and burnt another piece, and sowed oats directly, the crop of which was equal to the other.

Before Mr. Dupuis ventured thus upon a larger scale, he tried experimentally one acre and a quarter, the result of which was so advantageous, that he had little doubt of the success which attended his greater exertions.

About Caversfield, not common; some do it; but reckoned most profitable on furze lands.

The Rev. Mr. Plasket, of North Weston, pared and burnt 20 acres for turnip and rape; both good. He hoed part of the rape, and the stalks were so large that they were collected and burnt, and the ashes spread. The effect of which manuring was great. After those ashes, there was more corn than could stand.

At Stonesfield, upon the enclosure, part of a common of black loose soil, without stones, was, by Mr. Sotham, pared and burnt, and sowed with turnips, on one ploughing and an half. It was then worked well with a drag; crop fed off by sheep; ploughed once for barley, and laid down with ray-grass, trefoil, and red clover. The first crop mown, and the second year sheep-fed. Wheat put in on one ploughing; the crop good. Next, oats on one ploughing, and then the land thrown into the common rotation, beginning with turnips. From this, and many other trials, he is convinced of the great benefit, and even necessity of paring and burning, but thinks it has a tendency to waste the soil; and if repeated often, would be injurious.

Mr.

Mr. Taunton, at Ensham, has pared and burnt above 200 acres of furze land: a common waste enclosed by Act of Parliament. The furze being cut and grubbed, it was then pared and burnt, and admitted to be by far the best and most effective means of bringing such land into cultivation.

Mr. Pratt, of Fowler, has witnessed very many cases in which ploughing up sainfoin lays without burning, proved a ruinous practice; crop after crop have been eaten by worms; but Mr. Brook, the Duke of Marlborough's steward, would not permit it, to the great loss of the tenantry. Thanks to more knowledge, this is no longer the case.

Mr. Kimber, of Little Tew, who occupies his own land, and therefore certainly would do nothing to injure it, has a very high opinion of this husbandry, from 30 years experience. He is absolutely confident that it does not lessen the staple of the soil in the smallest degree, and considers that as impossible by this operation: but it converts all vegetable matter into the most valuable manure; it destroys grubs, worms, and insects, and it adds greatly to animal manure, by the quantity of stock kept on the large crops of every kind raised by means of it. He justly considers the prejudice often met with against this husbandry as an instance of great ignorance or great obstinacy, and has on all occasions contended in favour of it. He never practises it but for turnips, and never sows white corn the oftener on account of having done it; here he thinks is the error by which many have raised objections to it, from ill management alone.

At Enston, Westcot, Barton, &c. they pare and burn their sainfoin layers for turnips, which they think

far

far better than ploughing alone, for the land is in a hollow state, which always encourages *the slug*, whereas by sowing turnips, and eating them upon the land by sheep, the treading gives the firmness wanted. In order to keep the ashes on the surface, they half plough (rest-baulk) and then a clean earth.

The Rev. Mr. Filmer, Mr. Creek, and Mr. Wing, all agreed that this husbandry is excellent, and could not have had an opposer but from over-cropping, which does not of any possible necessity attach to it.

The common management is to plough twice, but Mr. Wing uses the skim-plough instead of it, which does the work far better. It is on the plan of Ducket's original trenching plough; a little iron plough in miniature fixed by two standard irons to the beam of the plough before the coulter. It will plough shallow, and much better than rest-baulking.

At Burford, a great deal of fresh land upon stone-brash; a twenty-acred piece pared and burnt by Mr. Turner, gave one ploughing for turnips; the crop good, and fed off with sheep. He then ploughed the land twice, and sowed a second crop of turnips of the tankard sort; crop excellent. They were eaten off early, and wheat sown, which was a great crop, and better than he ever had. Next barley was sown, the crop good, and laid down with white clover, which was seeded this year (1807). Of this husbandry of paring and burning, he has the highest opinion; and he would never break up sainfoin without it: has seen it done with mere ploughing for oats, and the crop eaten up by the grub. He thinks that if turnips are the first crop after this operation, no injury can be sustained : but if corn be sown, has some doubts.

Paring,

Paring, burning, and spreading the ashes, can be done on hollow land for 21*s.* to 25*s.* per acre : but in general it cost 30*s.* ; from 25*s.* to 35*s.*

Mr. Tuckwell, of Cignet, never breaks up sainfoin without paring and burning: he half ploughs it, and then a clean earth as thin as possible, to keep the ashes near the surface.

Mr. Pinnal, of Westall, could never think of breaking up sainfoin by any way but paring and burning. The difference in profit between that method and mere ploughing, is prodigious ; and is very sure that the land, instead of being injured by it, is rendered much the better for it. He has land of his own, which he has cultivated for eighteen years, which has had no other manuring than by this operation, and it is now in excellent order, and yielded this year (1807) a great crop of wheat ; yet he has never failed paring and burning it once in six years. All seeds upon such soils as are free from stones, are pared and burnt, particularly down or heathy land. He gives but one clean earth, as shallow as possible, for the turnips.

All seeds which lie two years are pared and burnt, provided the soil be down or heathy land. Would do the same upon the most brashy soil, if possible ; but the stones prevent it. Sainfoin is pared and burnt in March.

Mr. Edmonds is a great and decided friend to this husbandry, in every case where usually practised ; but he never does it where red clover with trefoil and white Dutch are sown, without ray, though it be left two years. But perhaps it is less advantageous on strong clays than on any others. First crop turnips, and second on old pasture ; and the first crop put in on

one

one earth of a skim plough, as called, but only common plough for going shallow. Never saw any instance of land that was hurt by paring and burning: instances of bad management have occurred in cropping, but in no instance by reason of the burning. He is of opinion that sainfoin never ought to be broke up without paring and burning.

From observation he knows, that on the thinnest soils on brows of hills, where repeatedly done in 30 years, it has not lessened the staple in the least. Heaps should be small; burnt as soon as possible, and not in too fine weather, to avoid over-burning.

He knows not one practical farmer who does not do it on his *own* land. They attend very much to ploughing shallow, not to bury the ashes.

Mr. Edmonds remarked, that if it was not for sainfoin, paring and burning, and turnips, the Cotswold farmers of Gloucestershire, and the stonebrash ones of Oxfordshire, could not pay one-third of their present rents.

Breast-Ploughing.—They have in various parts of this county a practice, which is certainly a very extraordinary one; that of paring stubbles as they pare grass for burning. This is in order to cleanse from couch, &c.; but as the depth they cut is not more than an inch, or two at most, it is an extraordinary fact, if it has the asserted effect of cleaning: yet one would suppose that farmers would not give 15s. an acre (the price of the operation), if the effect did not answer.

" The practice of paring and burning is in use in the part of the county bordering on Gloucestershire,

but

but is not very general. Opinions differ, in regard to
the utility and effect of this process. On light shallow
lands, the practice is said to destroy a portion of the
soil, already too scanty and weak; and though it
gives an unnatural and forced exertion for a short
time, it renders the soil much worse in future. In lays
of sainfoin worn out, and in rough pastures broken up,
where the soil is not too shallow, it is often found beneficial, especially if followed by a crop of turnips.
But great care should be taken, that none of the soil
be *entirely* burnt with the trumpery and weeds; because in such particles of the soil as are burnt so much
as to become red, vegetation is destroyed; and this
will be apt to happen, if the wind being high, the fire
burns too fiercely : but when what is left after burning
is only black, the principle of vegetation is supposed
to remain. It has therefore been recommended to light
up the heaps in an evening, as at times the wind is
often observed to sink, particularly in the months of
March and April, the common season for this operation*."

"These observations on this practice, are evidently
founded on the effect of bad management of the land
which has been burnt. The fact is, that the thinnest
soils are most improved ; and the deeper the land is
ploughed and burnt, the better."—*J. Boys.*

From all the information I received in this county,
I am confirmed in the opinion I have entertained, that
paring and burning is the best and most profitable
method of breaking up all sorts of grass on all sorts of
soil ; and that the apprehension of injuring the land is
absurd, and utterly unfounded in fact. Injury, and

---

* Original Report.

material

material injury, often ensues ; but it is from the cropping, and not from the burning : those who will not
make this distinction, must continue in their opinions ;
it is only to be hoped, that they will not have the management of great estates. If I made a tenant a present
of a great dunghill, I should not a whit the more permit him to take two crops of white corn running ;
why, then, is he to be permitted to take two, three, or
perhaps four, after burning ?

———◆———

SECT. III.—MANURING.

UNDER this head we have to enter intelligence received on,

I. Dung,
II. Pigeons' dung,
III. Lime,
IV. Ashes,
V. Rags,
VI. Mud,
VII. Gypsum,
VIII. Composts.

§ I. *Yard and Stable Dung.*—Mr. Daniel Percey,
in discourse upon manures, made an observation which
well deserves noting : *The dung of straw-fed cows is
good for little ; when yard-dung consists only of rotten straw and such dung, I think it hardly worth
carrying out.* Combine this with the repeated opinions of the Norfolk farmers, that they would not have
a handful of their straw eaten, but all trodden into

dung

dung by well-fed beasts, and then let us further reflect on the usual system of working straw-fed oxen ! !

Mr. Simmons, of Stokenchurch, and many other farmers, carry their stable and yard dung, long and unstirred, for strong land, and on that soil reckon it much better, load for load, than rotten dung.

At Waterstock, rotten dung is generally used ; but when laid on fallows, long, by Mr. Ashurst.

At Thumley, Mr. Rippington lays, like his neighbours, his dung in heaps ; " *mined*," it is called. They lay it on fallows ; but some directly from the yard, which, however, is rotten. They are attentive, when they plough it in, to let it be with a shallow furrow, lest it should be buried.

Mr. Bonner, of Bensington, throws his yard-dung on to heaps in the yard, if to be carried to any short distance ; but if further, carts it to heaps, and turns. Lays it rotten on layers for wheat : for turnips, however, he is of a different opinion, and if carried from Christmas to Candlemas, does not care how long it is, and is thus reckoned to be the best ; but would be bad for wheat, as it would keep the land too hollow. I may observe, that this reasoning is fair ; but be it remembered, that no dung should be laid on for wheat ; nor can it be, if the farmer has as many turnips as he ought to have.

Walking into a field from that excellent inn at Bensington, the White Hart, I found a man spreading dung upon a clover-lay for wheat, in so long and strawy a state, that it seemed fresh from the yards ; and the information I received was, that it belonged to Mr. Newbury, who, from being a grocer, was turned farmer, and had for a few years been in the practice of using all his dung in as long a state as the

season

season of using it would permit, and that he had found it more beneficial than keeping it with any view of rotting. Mr. Shrubb, at the White Hart, was the origin of the practice here, or rather his father; for it has been long the husbandry with them, and persisted in contrary to the general practice, from repeated conviction that it is *best* in that state, besides the saving in quantity, when compared with that which is rotten. He shewed me several pieces of clover land, with the dung spread quite in a straw state; much of it as long as if not made a week: also a very large field, and others smaller, where it was ploughed in; and one partly turned in, and part unploughed, with the straw on the surface, shewing clearly the degree of burying given by the common plough of the country, without skim, or other variation. It was extremely well covered, and where harrowed, very little indeed drawn out; yet the depth did not exceed four inches. It is to be noted, for it answers many questions, that the soil is a very loose, friable, gravelly loam, on sharp gravelly subsoil—land which any practical man would, on viewing, assert to be as unfit for this husbandry (if the system itself is or can be bad) as any soil that could be found. I questioned Mr. Shrubb most particularly on this practice, at first pretending much doubt of its propriety; he assured me that he had done it for so many years, and on so many very different occasions, and with so uniform a success, that he had not the smallest doubt of its profit. His crops, by means of it, have been so large (once five quarters of wheat per acre over a very large field, and at all times good), that he has every reason to believe that, load for load, it is better than rotten dung, which he thinks, by moving, turning, and rotting, loses its

most

most beneficial principles; in addition to which, he
has the advantage in quantity in a great superiority:
and further, that it lasts much longer in the land. In
order to render the land tight and firm, he pens his
sheep immediately after sowing and harrowing the
wheat; and drives the sheep several times up and down
the lands, in order for treading it the better.

Mr. Shrubb is a sensible observing man; and his
knowledge is not the experience of a small farm, on
which particular attention may command a certain
species of success not attainable on a large scale, for he
is one of the greatest farmers in Oxfordshire. At least
nine in ten of those I have conversed with in this county
have declared with peremptory decision, when I have
put the question, against the practice. A man from a
distance, and especially one who *deals in book husbandry*, may not be listened to; but here is the practice at your own doors. I beseech you, gentlemen, to
go on carting, turning, mixing, *mining*, and rotting;
give your manure to the sun and to the winds; continue to expend no trifling sums in the reduction of
four to one, in order, by studied operations, to render
your one less valuable than the fourth of the original
four: it is a wise conduct, therefore stick close to it,
and argue strenuously for it over the next bottle you
drink. *Mine* away; see that the heap lies light;
keep the carts off; take care that the air pervades it,
and let the sun shine and the winds blow—Who's
afraid?

Mr. Newton, of Crowmarsh, approves more of carting his dung to form mines, than throwing it up
in the yard. For beans, he forms the mine a month
before using; but does not turn it. He finds manuring

so necessary upon his soil, that he does it twice in four years.

The Bishop of Durham, at Mungwell, keeps his dung till it is quite rotten, and all fermentation over, whether it is his own made at home, or that bought at the town of Wallingford.

The farmer who occupied the farm before his Lordship had it, used long dung; and by that means filled the land so full of weeds, that the practice has been carefully avoided ever since.

Mr. Dean, of English, never mines his dung, as much is lost by that practice. He was spreading long dung upon clover for wheat, when I viewed his farm.

At Maple Durham, and its vicinity, they dung on clover lays for wheat: spreading that which is quite rotten.

Messrs. Foster and Hairbottle both agree upon the question of dung; they use only that *prepared*, or rotted, laid on heaps: the carts not being suffered to go in, but left as light as possible to ferment the better; and turned over once or twice. The idea of using long dung utterly erroneous; of which opinion Mr. Hairbottle had last year a strong confirmation: he prepared a field of sixteen acres for turnips, and having formed two dunghills at different places in the field, ready for depositing in the furrows of Northumberland turnip ridges, ploughed except where the dunghills prevented it, and the dung being applied, those spaces were then ridged and manured with long dung from the yard, and the field finished : the result was, that the crop was very superior after the first operation, to what it was after the second. The rotten dung answered greatly, and that which was long, was

much

much inferior : a comparison which he considered as
satisfactory. It is impossible to assert, that in this
particular case the rotten dung might be the better ;
it *may* prove better in various particular cases ; it is
at least possible. But there are many points necessary to
the *proof*, that were utterly neglected in this trial ; but
two essential ones are sufficient to note. The time of
both giving the tillage and of drilling, was different ;
a circumstance alone sufficient to destroy the authority
of the trial with a plant so much affected as turnips,
by these circumstances; and which would in many
seasons cause the difference of crop or no crop. Next,
how do we know, or how could Mr. Hairbottle know,
that the dung was exactly the same ; that is, made
by the same animals? The dung made in a farm-
yard, into which cow-houses, hog-sties, and stables
are cleared, may vary from 5*s.* value per load to 20*s.*;
and very considerably, even from the same stock at different times fed in a different manner. Such experiments and observations founded on them are only
sources of error. There is demanded in a comparative experiment, an attention to so many circumstances, every one of which must be strictly uniform,
that if the least be omitted, the result is affected, and
the experimenter only deceived. A general practice
for some years on a farm steadily watched, is more
valuable than particular trials, unless made with every
precaution.

The theory of Mr. Foster, is, that fermentation in
a dunghill deprives the heap of nothing that is good,
but adds something that is excellent, as fermentation
in malting barley, turns something that was inert into
spirit. When comparisons agree in one point and dis-
agree in an hundred others, the mind may, if it pleases,

fix on the one and overlook the hundred. If the fermentation of straws in a dunghill, were a vegetating process like that of malting of grain, and vegetation having effected the conversion that takes place, these straws were dried on a kiln, the case would not be parallel for want of many other circumstances; but it would come nearer to the purpose for which it is quoted.

Mr. Davy forms his yard-dung into heaps, not to rot it into spit dung, but for use when in fermentation.

Mr. Thomas Latham, of Clifton, applies the dung made from Christmas to Easter, for the turnips of the following summer. He throws it up in heaps in the yard; and if not, draws it out. What is made in the standing-fold is carted out often, to prevent the heat hurting the sheep. The heaps are turned once if necessary, but he does not conceive that the manure is better from being over rotten, as in that case it suffers both in quantity and quality. In regard to ploughing it in when spread for beans, it is done at a full pitch; for they conceive that as beans root deep, deep tillage is necessary for them. And though it is a maxim, that dung may easily be buried, still they do not regard this circumstance as sufficient to induce shallow tillage. When the bean stubble is ploughed for wheat, they also go a full pitch for the purpose of bringing up the dung before ploughed in for the beans.

Dung is formed into heaps about Baldon, in the manner common every where, and the prejudice of nineteen-twentieths of the farmers is in favour of its being very rotten. Sir C. Willoughby has carried it on some occasions long, to the land, and found that it answered very well.

Mr. Cozins, of Golder, *mines* his dung, and as soon

as the fermentation is in full height, carries it to the land.

Mr. Kench, of Enstone, gives all his dung to the turnip land; carries it out of the yard, and forms *mines*, which he turns once or twice to make it quite rotten: he has tried it in a long state, but it did not answer equally. If a crop of turnips rot with the frost, they are of very great service to the land.

The town dunghills at Chipping Norton are sold by auction, and such is the spirit of buying, that parcels have been sold as high as 20*s.* a cart-load on the spot.

Mr. Kimber, of Little Tew, does not approve of using long at all, in any case.

Mr. Sotham, of Stonesfield, cannot get his dung too rotten for turnips, and never heard any one of a different opinion. It is universal through Oxfordshire: mixing it with the scrapings of limestone roads is excellent. His farm, half stonebrash, and half hollow land.

Mr. Pratt, of Fowley, in common with his neighbours, lays all his dung on for turnips; carries it out to form *mines*, and turns over twice; some slovens will not bestow this labour, but such practice is bad.

Mr. Tuckwell, near Burford, lays all his dung on his turnip land; all made in summer, and the following winter it is spread for the turnips of the succeeding summer; and much attention paid to its being as rotten as the time will admit, by turning and mixing, as *green* dung is thought to be pernicious.

Mr. Pinnal, of Westall, lays all his dung on for turnips, mixed with earth or road dirt, and turned till rotten. He is very sure that long fresh dung would do more harm than good.

Mr.

Mr. John Tuckwell, of Burford, uses dung neither fresh nor rotten : he thinks it best in about three or four months.

Fresh dung, at Great Tew, not reckoned so good as rotten by many degrees.

Mr. Tiddmass, mixes the dung of his dairy cows with lime and bank earth in *mines*, turning over till rotten : will admit nothing in favour of long dung.

In the district of Thame, rotten dung is chiefly used ; but the Rev. Mr. Plaskett, of North Weston, prefers long dung, which, spread on grass upon cold land, forces a much quicker vegetation than that which is rotten ; and he finds, that on arable land it lasts much longer, at the same time that it keeps the soil more open and porous.

Mr. Creek, of Aston, does not approve of keeping dung in the manner generally practised in Oxfordshire : he turns it up in heaps in the yard in May, and in three weeks carts it at once on to the land, spreads and ploughs it in for turnips. Mr. Wing agreed in the observation, and is in the same practice. Mr. Creek does not think dunging for wheat advisable on a stone-brash farm ; the turnips demand all.

At Atterbury, they used to be very fond of rotten dung ; but of late years, Mr. John Wilson has had some doubts upon the practice, and several farmers have used it longer than heretofore, and find it in this case more durable ; but all still cart it from the yard to heaps. Last spring, Mr. John Wilson compared long and rotten dung upon a meadow ; and what was the result, Mr. Wilson ? *Why the long pleased me much better than the short.*

Mr. Davis, of Bloxham, lays his dung, the latter end of October, on the seeds of the spring corn crops:

not

not on turnips, because it makes the barley so great, that it destroyed the grass-seeds two years together. His seeds carried nine couples per acre all last summer, and kept them well : made the winter preceding, drawn to heap, and thrown up; not carted out till the dry weather was over, to prevent exhaling. Prefers rotten dung; never tried long fresh dung, but has seen it, and thought it very absurd : has seen on his own farm long spread accidentally, and bad.

Mr. Thos. Payne, of Drayton, lays his dung on for winter vetches, cabbages, turnips, and Swedes, generally at three or four months old : makes heaps in the yard, or carts out to make them, unloading by the side of the hill, and throwing it up, to avoid treading.

Mr. Singleton, of Bampton, thinks rotten dung best for grass land; but for arable (and especially if the soil be clay), long is much better.

Mr. Edmonds thinks that dung is, by many farmers, kept too long, for it is done to great loss of quantity : he would not put very long on for turnips ; but he has found that, on grass, to lay long dung directly after the scythe, has a great effect ; the worms draw it in, and it nurses up the grass strangely ; insomuch, that he thinks long dung nearly as good in that application, load for load, as rotten.

Messrs. Coburn and Secker, of Witney, both use dung (the former for wheat, the latter for turnips) of the last winter and summer ; not laid in heaps, unless it wants it ; and, in general, use it in a much longer state than common : they both think, that by keeping it till rotten, much is lost.

§ II. *Pigeons'*

§ II. *Pigeons' Dung.*

Sir C. Willoughby spreads this manure as a top-dressing for barley, and finds the use so great as to prove the benefit of a dove-house, were there no other object. In order to increase the quantity, he carries into the pigeon-house the poultry dung, and seeds of weeds from the barns (the heat killing all). The effect of the dressing is seen to an inch.

§ III. *Lime.*

It is, in the estimation of Sir C. Willoughby, the best of all dressings for sand, and does very well on gravel ; he has tried and found it so, and intends to burn brick and lime together, by which means the bricks may be had at 28s. per 1000 instead of 40s. the present price.

Mr. Thos. Latham, of Clifton, has tried this manure mixed with mould, and has found that it does most good on hot gravelly land.

Mr. Cozins, of Golder, tried it very amply ; but could not even see in the crop where it was laid.

Mr. Sarney, near Henley, can burn at $3\frac{1}{2}d.$ a bushel ; but his kiln is close to the wood : 65 quarters, in a flame kiln, for 10l. It has not done the good he expected. His father saw land limed in Kent, a bushel per square rod, with very great effect, which induced him to do it.

Mr. Fane harrowed in lime with the wheat-seed, and it did good ; but on a part he spread it on the wheat, and it killed the crop : but then harrowing in fresh seed, it proved good.

Mr. Taunton, at Ensham, has limed above 200

acres ;

acres; they were part of a furze common, enclosed by Act of Parliament. The soil, a yellow loam, part wet, and demanding hollow-drains—rarely a soil of natural fertility; but lime makes it very productive of oats, and prepares it for future wheat. The quantity, 20 to 30 quarters per acre.

Around Blenheim, and in the sphere of Mr. Pratt's observations, lime answers well on new land, but not on old cultivation.

Mr. Turner, of Burford, has found lime very good on a heathy loam soil, and generally every where, except where the stone is dug. The stonebrash burns to lime. He uses 20 quarters per acre, at 2s. 3d. per quarter: burnt by himself, and near coals; but the average price must be reckoned 4s. It lasts three years well, and is so beneficial on the right soil, as to be seen to an inch. He last year spread some for turnips, and it was this year (1807) visible at a distance in the barley. Mixes it with earth and road dirt, which are all calcareous: 15 quarters per acre, with some dung and mould. Lays it on for turnips, and can see it every where, and also in the seeds, and means to go on with it. The price is 4s. a quarter, but can burn it for 2s. 6d.

Mr. Walter Wilson, of Atterbury, limes on the young seeds, from ten to twenty quarters per acre; ten quarters common, and mixed with earth and some dung, on the fine red sandy soil. Some lay it on for turnips: for whatever crop the benefit is great, and lasts many years.

Lime answers greatly on the red land, a loamy sand, at Bloxham: they lay on from 15 to 25 quarters per acre. Mr. Warrener does much, and finds it to answer well.

This

This manure Mr. T. Wyat, of Hanwell, finds of great use on the red lands of that district. He lays 15 quarters per acre in half-bushel heaps, which are covered with mould, and left to slack, and then spread for turnips : the same year he lays on ten loads per acre of dung ; but never mixes the lime with it previously, which he has found to be erroneous management.

The Rev. Mr. Dupuis, at Wendlebury, has tried the pounded limestone shovelled upon the turnpike road, and found it, in every application, an admirable manure. A neighbour spread it on fresh grubbed furze ground, and nothing could answer better.

§ IV. *Ashes.*

Mr. Fane, 12 bushels of peat-ashes on sainfoin, or 16 bushels of coal-ash.

Mr. Simmons, of Stokenchurch, 15 to 20 bushels of peat, or 40 bushels of coal-ash.

Several farmers agreed, that peat-ashes on sainfoin, clover, pease, and turnips, was of very great service ; but of none at all on ray-grass. The quantity, 12 bushels per acre ; coal-ashes, 30 bushels. Mr. Hayward, of Watlington, 25 bushels of peat-ashes. Mr. Goodchild, of Greenfield, 15 bushels. Mr. Percey, of Bolney, thinks 10 bushels enough ; and that a very small quantity is sufficient, provided they be equally sown, which is the grand point in their use.

Mr. Sarney, of Soundes, has sown 5000 bushels of peat-ashes in one year, and 3000 or 4000 per annum many times. The price, 8*d.* per bushel.

Sir C. Willoughby, 25 to 40 bushels of coal-ashes. This manure is largely used by Sir C. Willoughby ; and a trial he made, comparatively shewed that those

of

of the Newcastle and Wednesbury coals are of equal benefit—there was no difference. Wood-ashes far better, but cannot be obtained in large quantities.

Mr. Bonner, of Bensington, uses 15 bushels per acre of coal-ashes, or 10 or 12 of peat, for turnips and clover, and finds the benefit very great, especially when they are got pure, without adulteration. Peat-ash never lasts more than for one year, but coal-ashes are sometimes seen in the wheat which follows the clover. Coal-ashes are rather the best on clover, and peat on turnips.

Mr. Davy, of Dorchester, a very excellent farmer, sows ashes on his clover and trefoil crops ; 12 bushels of peat-ash (*home burnt*, as they term it, that is, not in heaps for sale, which are much adulterated) per acre are sufficient. Coal-ashes from London, a larger quantity.

Many peat-ashes are used about Whitchurch, for sainfoin, clover, and turnips : the effect is considerable, and they have been found beneficial against the fly in the turnip crop.

Mr. Cozins, of Golder, having a vein of peat upon his farm, digs and burns it into ashes. He lays 50 bushels per acre, and finds very great benefit from it on clover and natural grasses. Dung, he finds, upon grass land, will give quantity ; for which reason, where he has dunged the land last, he next manures it with ashes, which gives a finer crop.

At Stanlake, &c. near Witney, many farmers sow ashes on their clovers, both coal and peat ; and malt-dust on wheat and barley.

§ V. *Rags.*

§ V. *Rags.*

Mr. Bonner, of Bensington, buys rags from London, at 9*l*. per ton, to spread on clover land for wheat, besides some land-carriage : 6 cwt. per acre, yet they last only two years.

Mr. Shrubb, of Bensington, uses them on clover-lays for wheat ; 3 or 4 cwt. per acre, at 9*l*. 10*s*. per ton, and 10*s*. carriage ; and also for turnips, for which crop they are ploughed in before winter, as soon as wheat-sowing is over : if ploughed in at the time of sowing turnips, they will not work for that crop. He generally gives half a coat of rags, and half a coat of sheep-fold.

The Bishop of Durham, on the clover land not manured with dung, spreads rags, 7 cwt. per acre, at 10*l*. per ton, including carriage ; but the dung is the better dressing.

Mr. Dean, of English, uses rags for wheat ; he lays on half a coat of these, and half a coat of dung ; and this, he thinks, is the most beneficial application of this manure.

Mr. Kimber, of Crowell, buys many rags from London ; the price, 8*l*. 10*s*. per ton, and cost 10*l*. on the farm. He spreads 6 cwt. per acre on clover-lays for wheat, and finds that they last longer than any other manure.

Mr. Tuckwell, of Cignet, has used many rags from Witney, at 7*s*. 6*d*. to 8*s*. 6*d*. per cwt. He spreads 5 cwt. per acre : they generally last only one crop— sometimes for two, and are superior to any thing for wheat. They are ploughed in ; and if the season be

very

very dry do not answer so well. Do best on a summer fallow for that crop.

§ VI. *Mud.*

Mr. Hairbottle, of Henley, emptied the mud from a pond communicating with the Thames, and spread it for turnips: it did no good whatever.

§ VII. *Gypsum.*

Mr. Sarney, of Soundes, sowed six bushels of gypsum per acre, on clover, in March, and the effect immense.

Mr. Davenport, at Shirborn, in 1807 sowed it on sainfoin and clover, and it did no good whatever.

Mr. Fane examined a field of lucern in Kent, over most of which it was sown, but part without any : and the superiority of the manured was great.

§ VIII. *Composts.*

Mixens of dung and earth, &c. are quite the practice of the farmers south of Oxford; and reckoned excellent husbandry.

Mr. Sarney throws urine on to malt-dust, keeps it six months before using; and the effect, when spread on the land, very great indeed.

Mr. Fane makes an ash-hole between two necessaries, and the compost taken from it is of the most decisive benefit.

Mr. Sarney is clear that there is no good whatever in beech leaves as manure.

Mr. Davenport, at Shirborn, has found that beech
leaves

leaves mixed with the grass of mown lawns and kept some time, has been of decided benefit.

The general system of the county may be considered as that of making composts; the yard and stable dung is *mined* with earth, with road dirt, or with lime, and in some places with all three; and he is reckoned the best farmer who makes in this way the greatest exertions.

SECT. IV.——IRRIGATION.

No watered meadow in the county, but there might in various parts of it be a great deal. At Brightwell there might be much, but the millers will not permit any; and it is decided in this county, that a proprietor has no right to effect any operation on his own property that may tend to lessen the quantity of water, if there be any mills below him. This may be good in law, but it is horrible in politics: wind and steam will grind corn, and therefore a water-mill has little pretention to utility; and if it impedes irrigation, is a nuisance.

SECT. V.——GENERAL IMPROVEMENT.

Mr. Fane is clear in the opinion, that the Chiltern hills of this county produce half as much again as they did 30 years ago; and the great cause has been the increase of live stock by more turnips and artificial grass.

In

In the opinion of Sir Christopher Willoughby, the agriculture of his vicinity is so improved in 20 or 30 years, that there is now ten times as much live stock as there was at the former period.

Mr. Pratt, of Fowler, who has known the country many years, and has had much experience as a valuer of land and tithes, and has acted as a commissioner in several enclosures, is convinced from a variety of observations, that the improvements which have taken place in 30 years are prodigious, especially from enclosing turnips and sheep : and that the food for mankind wherever enclosures have taken place is fully doubled.

Mr. Kench, of Enstone, is clear that both cultivation and live stock, through all the stonebrash district, are very greatly improved in the last 20 years.

Mr. Davis has known the county well for more than 20 years, and the husbandry is incredibly improved in almost every particular : if you go into Banbury-market next Thursday, you may distinguish the farmers from enclosures from those from open fields ; quite a different sort of men ; the farmers as much changed as their husbandry—quite new men, in point of knowledge and ideas.

CHAP. XIII.

LIVE STOCK.

—◆—

SECT. I.—CATTLE.

MR. Fane, at Wormsley, keeps only Alderney cows; he has had experience of other sorts, but finds these more profitable than any: whether for milk, cream, or butter, in quantity and quality, proportioned to the food, they exceed whatever he has experienced. He sold one which he did not like, to a farmer, whose family was not a drain to the dairy, and she gave 12 lb. of butter per week. He has heard of one on good authority that gave 16 lb. Mr. Fane has 13 and a bull; I saw the bull at plough with three horses, and he outwalked and outdrew his fellow decidedly.

The Earl of Macclesfield has a dairy of Devon cows and a bull. There are two of his cows truly beautiful; strait and round in carcass, clean and thin in leg and horn: fine specimens of that beautiful breed.

Sir C. Willoughby has a very fine dairy of 19 short horned cows and a bull; they give much milk, and he finds the breed profitable, but must be well fed, and in a great measure kept at home in winter, as they cannot stand the cold. He has remarked from long and attentive practice, that no hay should ever heat in the stack that is intended for cows. It should come out as green as it goes in, and produces more milk.

Mr.

Mr. Cozins, of Golder, keeps about 100 cows, of which 12 are for the dairy, 54 for suckling, 10 drying off, and 40 fatting. The barren ones are fed on straw, but all the rest on hay; besides which, oil-cake or corn is given to the fatting ones; he has compared six on corn and hay against six on cake and hay, and from that and the result of other trials, concludes that the best corn feeding is to grind one sack of beans with three sacks of barley, and taking the beans at 36s. per quarter, and the barley at 35s. and cake is cheaper at 13l. 13s. per 1000. His breed is generally the long horned, as he finds his land too cold and wet for short horns, whose hides are too thin to bear it. He has tried Herefords, and do well. Mr. Cozins has a very favourable opportunity of comparing the profit of a dairy and of suckling; and he is convinced that the former (after deducting the extra expenses) is one-fifth more profitable than the latter. The calves go to the London market at 12 or 13 weeks old. He esteems it a good cow, that gives 6lb. of butter per week, for three months after calving.

Much rich land in the vicinity of Thame: at Worpsgrove, six miles from thence, letts from 30s. to 3l. per acre. Mr. Bonner pays 3l. for some under cows, but the landlord pays all taxes whatever: the best land summers a cow to an acre. Mr. Bonner prefers the long horned breed: he has tried the short horns, but found them too big and tender. Some Welch do well, being good milkers, but when fattened they are very deficient. In winter, all are fed on hay; and each cow eats 2cwt. per week, which he knows from having weighed it to them; while at hay they go in and out as they like it. In fattening he has given oil-cake at 14l. 14s. a 1000: if the cow was not fairly

saleable

saleable from other food at the usual time of getting off, and meat advances by keeping her 1*d*. per pound, oil-cake answers even at that price; but if no advance takes place, it does not answer. The manure is the grand object, for he is sure that one load of it is worth two of that from hay feeding, supposing litter not to be given too profusely. Their carts hold about 30 bushels, and for such a load he would give 10*s*. for oil-cake dung. He conceives that a cow on oil-cake (but never measured it) would make a load of dung per month; if so, the dung is worth 2*s*. 6*d*. per week.

The rich district of Thame is applied to the dairy, or to suckling: the breed, the short horns. None bred but all bought in calf at three years old. They are kept a year and a half or two years, or proportionably to their time of keeping in milk. The dairy is reckoned more profitable than suckling; but also more troublesome. Upon an average, the whole dairy through, and all in milk, will give 5lb. to 6lb. of butter per week: but here a very uncommon circumstance takes place; for this quantity will be produced the whole year through. The cows hold their milk a year and more. Mr. Plaskett, of North Weston, had a cow that held her milk six years, and never had a calf in that time, and at last gave two gallons a day. Hence, five or six pound of butter a week, which seems a small quantity at first sight, is in fact a great produce. These cows are fed upon hay only, in winter; and eat the produce of two acres, which yield two ton per acre at the first cutting: the second crop is usually given to sheep. In stocking these rich pastures, sheep are always mixed with cows; and five are reckoned equal to a cow: the breed, Wiltshire or Berks.

An

shire. An acre and a half will carry a cow and two sheep through the summer.

Some farmers fatten their cows on oil-cake, and for very large beasts it answers ; but hay is more general.

The Bishop of Durham, at Mungwell, has some very fine short-horned cows, Alderneys, Yorkshire polls, &c. He thinks, however, that the short-horns are too large a breed for the Thames meadows, which are occasionally flooded. And as these waters come from gravelly soils, they are not very rich.

Mr. Freeman, of Fawley-court, has 20 short-horned cows, 20 calves, 20 yearlings, and 20 two-year old, and 20 cows, or three-year old, fatting off; and works 20 Devon oxen. Here is a dislocation of a most excellent system, by not breeding the race that will work well : if such a stock is kept, assuredly all should be Hereford, or Devon, or Sussex, that working may be a part of the general system. But Messrs. Foster and Hairbottle do not approve of working oxen ; asserting, that with a Rotherham plough and a pair of horses, they can plough as much land as six oxen can do. The horses do three-fourths of an acre in autumn, and an acre in summer ; whereas four of Mr. Freeman's oxen do only half an acre a day; thus, in fact, it requires eight oxen to do an acre, for four having ploughed half an acre, are changed for four others to do the other half. I must confess I do not understand what this can mean, for the oxen are fed on grass and hay, and never work on straw. But the allowance of such practices is not the way to eradicate, but to establish prejudices.

At Waterstock, Mr. Ashurst has the grass lands under cows bought in, of the short-horned breed ; they are generally dairied, and the butter all sent by contract to London. They are fed in winter on hay ; none on straw except rough cattle. Cuddesdon is also under dairies, and many dozens per week sent to London. An acre and a half will summer a cow : the cows cost from 20*l*. to 25*l*. each, and are in general very healthy. Sheep are kept among all the dairy cows. Ewes bought in at Michaelmas ; the lambs sold fat in the spring, and the ewes also fat in the summer. They used to be of the Wiltshire breed, but they are changed by many now, to a cross between the Leicester and the Cotswold : there are, however, many Wiltshire still. In one or two places in this cow district, kept grass for spring use is known, but not at all a general practice. On farms which have turnip land, that root is very rarely given to cows.

At Thumley, near Waterperry, Mr. Rippington keeps both long and short horned. The latter give most milk, and improve most in value, but they are tenderer than the long-horned, which suit better on cold land. All are applied to making butter for London : the average produce may be reckoned 6lb. per week in summer. All are kept on hay in winter ; but if dry early, some are on straw, not many, nor for a long time. If any straw be eaten, 50 cwt. of hay will winter one : the summer food about two acres. He has, or knows a piece of 64 acres, which feeds 20 cows, besides sheep in couples ; the cross South Down and Wiltshire, very few Leicesters.

Dairies are found on a large scale at Wendlebury and Bicester, and they are reckoned to yield, the whole herd through, on an average 5lb. of butter per week. The breed long-horned ; the land not thought good enough for short-horned. All fed on hay in winter.

Mr. Forster, at Bignal, prefers the long-horned breed

on

on his stonebrash farm: he works five heifers spayed, which at seven years old fatten so well as to sell high; a pair of them have sold for 100*l.* His nephew informed me that he has himself sold a fat cow of this breed for 60*l.* and one for 45*l.* this summer (1807).

Much dairy ground in Hampton Poyle; Mr. Blake keeps 60 cows, Mr. Irons 40; Mr. Savoury near 40: he makes good cheese at 60*s.* cwt.; the other two butter for London.

Mr. Tiddmass had for many years chiefly long-horned cows for his dairy, but of late he has bred mostly short-horns, and prefers them; not so much on the comparison of milk and butter, as in respect of carcass coming to more weight. They give more milk, and though the common idea is, that the milk is not so rich, he is inclined to think that this depends in a measure on their not (in the cases that have given rise to the opinion) having been bred on the land on which they were afterwards milked, which he thinks a point of some importance. Such as he has bred himself, give milk as productive in butter as the long-horned breed. Many prefer the cross between both. He winter feeds on hay, and does not at all approve of straw feeding: the butter goes to London twice a week. The pigs are not bred, but bought in at four or five months old, and fattened by the milk, feeding six weeks or two months. He prefers having them calve all through the year, and not to come at any particular season. No dairy-maid in this country milks, nor in Bucks or Northampton; all done by men or lads; and over in an hour, or at most an hour and a half, whatever the number of cows. They are tied up for milking; and Mr. Tiddmass has converted an old shed into a house for tying up the cows for wintering

on

on hay, which is very conveniently contrived; there is a space for a path at their heads and behind the feeding manger for giving the hay, and windows in the wall for receiving the hay directly from the stacks, which are built for that purpose parallel with the shed, and but a few feet from it. On the other, or opposite side, are the doors, with spaces beside to open or close at pleasure, for regulating the warmth according to the weather. They are turned out for drinking and air, at night and morning. In this way of feeding, he is certain from experience, that the cows (especially if they are large beasts) do much better, and are supported with less hay than if abroad.

Mr. Bowles, of North Aston, has a dairy of 20 long-horned cows which have been bred or selected from Leicestershire with considerable attention, and are much to his credit. I found them in a great state of flesh, and one fatting, which was a good specimen of a disposition to be fat: these cows are in such a state as should seem to prove that Mr. Bowles (he was not at home himself) preferred the breed for the readiness with which they are at almost any time transferable to the butcher; for it is not often that such cows are good milkers. In the flow of the season it demands three of these cows to fill a four gallon pail; which we should in Suffolk consider as miserably bad milkers; but then our cows when dried, demand a large expense to fatten; and if sold lean, produce but a trifle. Twenty cows produce nine or ten dozen a week, besides supplying a large family: if there were no family, the cow-herd thinks the quantity would be eleven or twelve dozen at the season; some in full milk, others middling, some going off, and some perhaps dry: this may be called 7 lb. a cow at the highest;

and

and seems to confirm other accounts of 6 lb. and 5 lb. :
I should conjecture that the very best cows on an
average of the full season gave 7 lb. commonly good
ones 6 lb. and the mass 5 lb. Three breeding sows
produce pigs sufficient for the skim-milk: they are
partly fattened on it, and finished with barley-meal.
The cows are themselves, when fattened, finished with
ground barley to the amount of three quarters of a
peck per diem. Mr. Bowles never ties up unless for
fattening; they are out winter and summer at grass,
being brought home morning and evening for a sufficient time to eat their hay, which is the only winter
food.

Mr. Rowland, of Water Eaton, prefers the short-
horned Yorkshire cows, which not only give most
milk, but improve most after buying, and when fat
bring most money. Mrs. Rowland (sister to Mr.
Westcar, of Creslow) assured me that their milk is by
no means so inferior to that of the long-horned cows as
some persons had asserted. But Mr. Rowland is making a considerable trial of another breed of cattle, the
Herefords: having at different times a few cows of that
race in his dairy, he liked them sufficiently to induce
a larger experiment, and he has now 21 Hereford
heifers.

He has found the milk richer, but not in equal quantity with that of the short-horns; and he is inclined to
think, from his practice hitherto, that the mixed
breed between the long and the short horns, is better
than either of them pure. The first winter after buying in, the short-horns are apt not to do well, but
afterwards just as well as others. He buys at three-
year old, and they grow so much, that they ought
not to be fattened till they are seven or eight years
old. He fats them to fourteen score a quarter; generally

rally from twelve to fourteen. In winter they are fed upon hay and grass, and fattened off with some oil-cake. In regard to the time of buying in the oil-cake, he remarked that fresh cakes were as good again as old ones. One year he sowed a large piece with flax, to produce seed for fattening his cows: he ground it with pollard; a sack of pollard to a bushel of flax, and it answered very well; but still conceives it to be inferior to cake. At ten guineas a thousand, he would feed with cake rather than with hay.

Mr. Weyland, of Wood Eaton, keeps short-horned, which give more milk than any. He has a fine tract of meadow, which join those of Water Eaton, and therefore he can support these large cows; but they are found to be rather tender: the winter food is hay, and in summer, the best land will nearly carry a cow to an acre. Much manure is spread upon the grass land.

Waggons go from Bicester to London with butter, taking ten tons per week: chiefly loaded between Bicester and Wheatley.

Mr. Wyat, of Water Eaton, whose fine stock of cattle I viewed with much pleasure, prefers the short-horned breed, and has many very fine cows of it; from 60 to 70 of all sorts; he both buys and breeds; has a bull got by the sire of the famous Durham ox, and breeds bulls for sale: some of which promise well. He has many cows of the half breed between long and short horns, and this cross he seems to prefer. He has a few Hereford cows, but is not inclined to change his stock in favour of them. His cows in winter are fed entirely with hay; to which for fattening, oil-cake is added at times, according to price and other circumstances. Barley-meal he uses for the same purpose, and has also fed with flax seed, of which he has had

crops

crops for this purpose, and he considers flax seed at 9*s.* or 10*s.* per bushel, as cheaper than oil-cake at 15*l.* or 16*l.* per 1000 ; barley at 32*s.* or 33*s.* per quarter also cheaper than cake at 15*l.* He fattens the beasts entirely in stalls, never turning out ; and he thinks they do much better than when loose in yards. But his milch cows are out at grass all winter, being tied up in the morning and evening to eat their hay : floods are mischievous to him, and in such cases he is forced either to buy hay or lessen his stock.

At Bicester, long-horned ; in the country are many short-horned, and there are cows that give 12 lb. of butter per cow, for a month or six weeks, or more ; and a notion is current at present, that a middling cow will yield a produce of 15 lb.

There is much very good grazing land under cows at Atterbury, both for dairying and fattening. Mr. W. Wilson has had 100 in one year. They feed them on hay : asking Mr. John Wilson if ever fed on straw : *No ; straw be a good thing to lay on.*

Mr. Salmon, of Hardwick, keeps 80 cows, and has his three churns so attached to one standard of a water wheel, that they are all turned at once; and in case of frost, or a want of water, a horse-wheel works them. Twelve breeding sows produce pigs enough for the offal of these cows.

Mr. Salmon weans 20 calves every year, on flax seed boiled and mixed with milk and water, and also with oatmeal ; half a gallon of oatmeal boiled in half water and half milk, ten gallons a day ; the flax boiled by itself in water before, and a quart to this quantity : this weans them as well as possible, and will even fatten them. Ties up 50 cows out of 130 wintered beasts, every year ; hay goes much further, and no poaching :

those

those that are out, are in lots in fields, with a warm hovel in each, and a bin for hay, which is stacked by it ready, according to the number; this for store cattle: and he has found it highly profitable management. Has now 90 cows in calf: milks from 75 to 80; butter 47 dozen. A man milks ten cows in an hour and a half.

In summer, when the milk is very rich, half water to it, and a gallon of beans night and morning to ten pigs, and the best food of all. The pump so fixed as to pour at the same time into the cistern of milk.

Cows all come in from February to May, which is a great matter. Prefers long-horned cattle, but has a few short-horns; he breeds from Mr. Astley's, in Leicestershire, and has given high prices.

Mr. Davis, of Bloxham, long-horns, but getting into short, as giving more milk and more butter, and the general opinion in Bucks and Berks is, that the short-horn is the more profitable breed.

Mr. T. Wyat, of Hanwell, has some remarkably fine cows of the long-horned breed, which he prefers to the short-horned, as he has found that in feeding they eat less, as in the proportion of two to three. They do not give so much milk, but it is of a richer quality; and they fatten in a shorter time. He had this year both breeds fattening in the same grounds, and going together, and the long-horns are gone to market, but the others not yet ready. He milks them for the dairy: in winter the food is hay, but some have straw till within six or eight weeks of calving. He fattens some on cake for finishing, but not much: the price 7*l.* 10*s.* to 12*l.* per ton. The cakes made in Warwickshire are much thicker than the London ones, and he thinks better,

There

There is in the rich parish of Atterbury nearly as much grass as arable; and under dairies and fatting cows, both long and short horns, but in general the former, as the more hardy, and fatten better; but the short give more milk, though not so good a quality.

Mr. Coburn, of Witney, has built a very complete ox-house for stall-feeding: he fattens Herefords chiefly on hay only, and has sold as high as 36*l.* and 37*l.* per ox. The hay is cut into chaff, and about a sack of barley to finish each ox, ground into meal. The stalls, feeding, and water-troughs, with other conveniences, are very well executed.

If we shine, said Mr. Edmonds, in any thing, it is in good hay, and it fattens well; and the cattle he has shewn in Smithfield never ate any thing else: has not bought any cake for twelve years. The winter made cake is by far better than that made in summer; the latter is dry and hard.

Mr. Edmonds is decidedly of opinion, that it is much better to have them eat their food out of doors rather than within; there should be a current of air where the beast eats, which prevents tainting the food with his breath; he should be cool, but not exposed: single stalls the fore part of the winter, but when the sun warms, the open yard and sheds may do well. In feeding, hay should be given thrice a day, and it is of great consequence to give what they will eat and no more; over feeding bad, and the cause of many failures, though the waste be taken away.

Mr. Edmonds has on these ideas built a very capital range of stalls in the Hereford way, an apartment for each ox, but the manger (deep to the ground) on the side opposite to the shed, and there are the hay stacks

for

for feeding. Each stall 9 feet wide, the house 170 feet long, 10 feet wide, and the yard 12 feet.

Mr. Singleton, of Bampton, who grazes largely, buys Hereford oxen in the spring, and some in the autumn: the former are turned to grass, and in November to hay and bean-meal, or that of barley, but beans better: gives half a peck twice a day, hay at the same time. When beans are 7s. or 8s. per bushel, or barley at 6s. would not give either. He thinks cake at 14l. or 15l. per 1000, too high; but at 13l. uses it. He considers cake at 13l. and beans at 6s. 9d. about on a par, but that if there be a difference, the beans are the cheaper food: the beasts, however, come on faster on cake. These articles of food will not answer without the expectation of 1d. per lb. rise in the market price of beef. Many of his own oxen are of a size that demands such feeding, as large beasts will not fatten so soon as smaller. The largest beasts will fatten on good hay, but they take more time; in which case they eat each two cwt. and a half per week, but if they have cake or meal, less: beans save hay more than cake. I viewed his numerous stalls and yards with pleasure, and the hay-stacks well disposed for feeding without trouble or waste. He ties up some, and has others (especially the larger beasts) loose in open stalls in the Hereford system; and these he prefers, as the beasts travel better: they are attended also with less trouble, and they *do* rather better. Oxen tied up make a load of dung per month: upon inquiring the value, he thought 10s. too high. Observing a small dunghill squared up by itself, and inquiring concerning the beasts that made it, it seemed to result that a beast made near a cubical yard rotten per week, which does

not

not ill agree with the former idea of a load of long per month, or the quantity decreasing in about ten or twelve months, three-fourths: with plenty of litter would make much more. He has yards for four or five beasts, but objects to them, as there will always be a master that thrives better than the rest. Cows are better to tie up, from the fall of the urine being clear of the bed. In hay feeding, they find that such as heated in the stack loses much of its *proof*.

The short-horned cows Mr. Singleton considers as much more tender than others; their thin skins expose them to flies in summer and to the cold in winter.

In all stall-feeding, Mr. Singleton has no doubt that keeping the beasts hot is bad; they should not, however, be cold, only cool.

The intelligence relative to cattle in Oxfordshire, is not locally interesting. The county has no breed of its own; nor is any particular race so predominant as to afford much information that is particularly valuable; much the greater part of the county is arable; and in the very narrow districts where grass prevails, there is not much to excite the attention of the traveller.

SECT. II.—THE TEAMS.

§ I. *Horses.*

Mr. Cozins, of Golder, in common with so many good farmers in Oxfordshire, never permits his horses to rest in the stable at night; they are always turned into a yard he made for that purpose, along one side

of

of which is a range of sheds for them to retire under, provided with racks and mangers; and a bin against the opposite fence, in case any younger horses should be mastered by the older ones. This system he holds to be of much consequence in keeping the teams healthy.

Mr. Thomas Latham, of Clifton, considers malted barley to the degree of just sprouting, much better than oats for feeding horses in the spring: four bushels as good as six of oats. He once fed a horse which he kept for another person, attentively; the food being weighed and measured, and though not fed higher than common, it costs something more than 25*l.* for food only. In order to lessen the expense of his teams, he buys in at three, four, or five years old; and after working two or three years, sells them to some profit.

In common with all other good farmers in Oxfordshire, his horses lay out in a yard every night: Mrs. Latham, his mother, who once farmed above a thousand acres of land, but gave several of the farms up to her sons, has a very excellent yard for this purpose, with water, sheds, and every convenience: it is one of the best yards I have seen in Oxfordshire.

At Atterbury, they tie their horses by the leg to feed clover and vetches, but Mr. W. Wilson does not approve of it; and never practises it; nor does Mr. Bellow ever do it, as he much dislikes it.

Mr. Pinnal, of Westall, near Burford, and in general all the farmers in that neighbourhood, make it a rule to have all their teams lie out in the yards every night. Some years ago it was the custom for them to be kept in the stable, but the present practice is found to keep the horses in a much more healthy state: they have sheds to go under.

At

At Stokenchurch, an arable farm of 360 acres, twelve horses.

Soiling is practised in the part of the county south of Oxford for horses: Mr. Percey and Mr. Hayward reckon sainfoin excellent for this purpose, but clover is bad; nor are vetches good except for a very short time, and when advanced in podding they are not wholesome: in a word, nothing better than hay all the year. Every thing I met with about soiling, was in a tone of voice and manner which made me suspect that they have not thoroughly made up their minds to the practice; the Oxfordshire farmers speak of it rather coldly: they are decided in condemning clover for this purpose; and have a notion that it gives horses the staggers.

Mr. Newton, of Crowmarsh, does not soil his horses, as he thinks it robs the land too much, but feeds seeds and vetches with sheep.

I had much hope, when I heard of two Northumberland farmers being brought by Mr. Freeman, tc Henley, that the soiling system would be much extended, and great quantities of dung raised through the summer, as well as in winter; but from the conversation I had the pleasure of having with them afterwards, I conjecture that less stress is laid upon it in Northumberland than in several of our southern counties.

In the Dorchester district, winter tares are universally sown with a view to soiling the teams; given both in the stable and yard: nothing reckoned so wholesome. It is common to sell them from 1s. to 1s. 6d. per pole, and always attend to clear a land at a time, in order that the ploughs may come in immediately to turn up the land for turnips, which succeed extremely well. If the land was dunged well for winter wheat, the tares succeed without any more manure, unless sometimes some

some coal ashes in the spring; but if it was not dunged, then they dung for vetc..es: they do not soil upon clover, considering it as very unwholesome for them. Mr. Latham knows a farmer who has his team all distempered by this practice: some blind, others swelled legs, and others broken winded: it was from second cropped clover soiled in September. In order that the soiling upon tares may last the longer, they are sown at several times, and spring tares also to succeed the others. They have an idea that much water is unwholesome for horses, and therefore are very careful in letting them have it.

Mr. Bonner, of Bensington, soils his teams on vetches, and never turns them out: some do it on clover, but vetches considered as much better, and healthier for the horses, as the former is apt to fill them with humours, and even to affect their sight: he thinks that horses must have a sound constitution to stand long soiling on clover.

Mr. Costar, of Oxford, who keeps above 100 coach horses, soils them on vetches; they are in summer racked up with them, and always do well: none can be in better condition.

Mr. Turner, of Burford, mows vetches for soiling his teams of horses and oxen.

Mr. Pinnal and Mr. Bagnall have a high opinion of this practice, both for horses and oxen. Mow vetches and some sainfoin from about May-day, and keep on until Michaelmas. By this means they raise great quantities of dung, and are very sensible of the general importance of the practice.

Mr. Walter Wilson, of Atterbury, sows vetches for soiling his horses in the yard, where he has sheds for them to go under at pleasure

§ II. *Oxen.*

§ II. *Oxen.*

The Earl of Macclesfield, at Shirborn, works three
or four teams of Devonshire oxen : his Lordship assured
me that he has not the smallest doubt of their being
more profitable than horses ; and especially in the vale,
where he finds shoeing quite unnecessary ; they work
well on hay, and never have corn: four in a plough,
and do as well as horses.

Sir C. Willoughby has great doubt of the profit of
using oxen, and he founds it upon the practice of the
farmers. Lord Harcourt keeps a team of Herefords,
and within fifteen years there were seven or eight farmers in the neighbourhood who kept and worked oxen,
but all of these have given up the practice. Now it
is contended, that in an article free from all ridicule
among one another, and the practical circumstances
of the business well known, with men ready to work
them, their being abandoned proves that the practice
cannot have any extraordinary profit attached to it.
It is to be considered also, that Oxfordshire has no
breed of its own ; it is not a breeding country, except
for suckling veal, which answers here, it is supposed,
much better. How far this may account for the fact,
is not insisted, but so it is, and the benefit must be
proved clearer than it has been, before he will adopt it.

Mr. Cherrill works a team of four Hereford oxen,
which do as much work as three horses : no other ox-
team in the district, except Lord Harcourt's, who
uses two teams. Seven years ago Mr. Thomas Latham, of Clifton, had a team of four, that drew with
ease ten quarters of wheat in a waggon, and were far
beyond

beyond horses for timber carting; they were Scotch beasts, and cost 36*l.* the four. He worked them three years, and sold them lean for 48*l.* They ploughed as well and as much as horses, but did not cost so much by any means.

Mr. Lowndes works oxen at Brightwell; he has seven Devonshires: they are fed with hay, chaff, and bran, and have a few ground oats: they plough as well and as fast as horses. The quantity of dung made by them, is double that made by the same number of horses in the manner they are kept by farmers; is sooner operative on the land, and more beneficial to land of the quality of his farm.

Mr. Dean, of English, tried working with oxen, but it would not do.

Mr. Cozins, at Golder, never worked oxen, but has seen them often, and thinks them so inactive that they cannot work with horses. *Have you seen the teams around Burford?* I have not.

Mr. Forster, at Bignal, works a team of five spayed heifers in harness, of the long-horned breed: he begins to plough them at two years old, are in full work at three, and they are fattened at seven: he has sold them as high as 100*l.* per pair. They are not shod, though on this stonebrash surface work as well as horses; and Mr. Forster has mentioned to his nephew an intention of having more of them: they work better, he thinks, than oxen.

The Duke of Marlborough has one team of oxen, Glamorgans; four in a plough, and work to the full as well, and plough as much as horses, and walk to the full as fast: more profitable than horses.—*Mr. Palmer.*

Mr.

Mr. Turner, of Shipton, has two ox-teams (five each) at plough : they plough an acre a day, do their work very steadily, and as readily as the horse-teams. He begins working them at two years old, and works till they are five. They live upon lattermath ; when that is done they have a little hay, and lie in the straw-yard, eating straw in the night. When they leave working he puts them into lattermath, then to hay and turnips, to fatten : upon an average sells them at 40*l.* an ox.

Mr. Preedy, at Great Tew, has one team of four long horned ; and Colonel Cox has another : they are considered as slower than horses, but do not cost near so much, nor require the same attendance when their work is over.

Mr. Kench, of Enstone, has five teams of Hereford oxen, which he buys in that county at three years old, and works for two, three, or four years, just according as *times* favour the sale of them to graziers : they work as well (five in a plough), and walk as quick as horses. He works them while fed on grass or hay, and sometimes on hay only : nothing better for them than sainfoin hay. He has been in the practice many years, and has not the least doubt or question of their being more profitable than horses for home work, but on the road uses horses. He never shoes them ; he buys them at 10*l.* or 12*l.* each, and after working them three or four years, sells them at 16*l.* or 18*l.* ; but for horses he gives 20*l.* or 21*l.* Gives only a bushel of oats per week, racks them up at some seasons with pea or bean straw, yet oxen far, very far cheaper Let it be remarked, that Mr. Kench is not a whimsical *gentleman,* but a common farmer, that pays rent for above a thousand acres of land, has a very clear head, and a good understanding and long experience : and may I not ask after this, whether any impertinence can be like that of men's condemning the use of oxen from theory and calculation, who never worked a score of them in the course of their lives?

Mr. Rowland, of Water Eaton, works two teams, and is clearly of opinion that two oxen are cheaper kept than one horse. Four oxen plough as much and as well as four horses: walk as fast, and upon the whole, have *rather the best of it :* all Herefords, and will draw any thing. If worked, they are fed upon hay, but if not worked, are at straw. He is very sure that an ox which has worked, will fatten far better than one that has not worked: they should never be fattened till six years old, nor is there any objection to working them till seven or eight years old. And Mr. Wyat, of the same place, would use them if he had arable land; but has only a few acres for turnips.

Mr. Jones, of Islip, and Mr. Whorrod, both work oxen.

Mr. Turner, of Burford, has three teams of oxen, of the Glamorgan breed: puts them to work at three years old: works them two or three years, and some-times four; then fattens them on rouen, and hay, and turnips, finishing with Swedes.

Mr. Tuckwell, of Cignet, has always one or two teams: I walked thither from Burford to see them work; a horse and an ox plough were following each other in the same furrow, four horses and four oxen, and the oxen trod on the horse ploughman's heels, step for step, just as fast as the horses, walking with ease. I then went to his house, and he readily answered my inquiries. The breed is Hereford, bought in at three years old, but some at four or five, as they may happen to be procured. He works them at two or three years,

but

but thinks that they should not be worked beyond seven, or at most, eight years of age : the price he gives, from 18*l.* to 24*l.* : sometimes has sold them to graziers, but in general feeds them himself; yet has scarcely any permanent grass, as his is a stonebrash arable farm. They eat sainfoin, the grass seeds of the course, or vetches; and in working, thinks it of much consequence that they should always have a little dry meat with green food : some hay cut into chaff night and morning. The same for fattening, with turnips sliced into it, and finishing with Swedes. The fatting begins about the middle of July, and they are finished about the middle of April; but as sainfoin hay then gets a little harsh, he gives, late in the fatting, a little barley meal.

In regard to the benefit of working them, he could not readily conceive how any one could doubt it (note, it is the common husbandry around Burford, almost every man having them); that they are much more profitable than horses, he has not the shadow of a doubt: to keep one team of horses is useful, but all the rest should be oxen. Whenever they are not found useful, he thinks, from all he has observed, that the reason is their being improperly fed and driven. To make it so cheap a scheme as to work on straw, or any oxen that are not in good flesh, is the sure way to fail; they should be so well fed at all times, worked or not worked, as to be kept in good flesh; if they were always full half fat, it would be so much the better; they then are in heart, will work without losing flesh, and are always ready to fatten in proper time : an ox should be fattened not so much by change of food, as by merely resting from labour. Thus managed, they are as strong as any horses, and will work just as well;

or

or at the most, the difference is not more than as four horses to five oxen : but with him, four oxen have all this summer done more work than four horses. To turn them to straw because they do not work for a month or two, he holds to be very unprofitable, and while at work, they should never lose flesh, and always thrive though worked. To let them go back in winter, and feed just when they work, is utterly unprofitable. The worst food he gives when they do not work, is cut straw, with a mixture of ordinary hay.

If they are hard worked in barley sowing, he gives them a little barley meal night and morning, the quantity small; but at all events does not let them lose flesh, as that is much more unprofitable than improving their food. Driving is another object of much consequence ; they want more attention than horses to keep them equally in work. All plough at length, both horses and oxen, and in harness.

In health, and general freedom from disease, they are superior to horses ; he does not even recollect having a lame ox.

If it is asked, if oxen be thus beneficial, how come they to have been given up in so many districts ? the answer is, we know not; for at and around Burford it is just the reverse: where one ox was kept ten years ago, there are now at least ten ; and they spread every year: that they are more profitable than horses he has not the smallest doubt : it is not a matter of reasoning in his mind, but practical conviction. When the ox is sold, it is (whether to a grazier or kept for fatting) to profit : how is this with a horse? Who does not know the expense of horses in the decline of value? In a word, the ox, after doing as much work as the horse, is sold to

profit;

profit; the horse to loss. Where the doubts are to be found, I know not (added Mr. Turner), but assuredly they will not be found at Burford.

Mr. Pinnal, of Westall, near Burford, one of the greatest farmers in the county, keeps more oxen than horses, and has one farm of 300 acres without a horse upon it; and he has not the least doubt, but that oxen are, upon the whole, as cheap again as horses. Upon a farm at Westall, of 600 acres, he keeps ten horses and twelve oxen: the country is all stonebrash, and upon above 2000 acres, he and Mr. Bagnall have not more than twelve acres of meadow; so that the common idea, that oxen can only be kept profitably where there is much good grass land, is completely refuted by the practice of this great and well managed farm. They keep between fifty and sixty Hereford oxen, which is the breed they prefer; have tried the Glamorgan, also the Devonshire, but these are too light: use five in a plough (sometimes four), or four horses: never shoe them: begins to work at three years old; purchased at Ross or Hereford, at from 15*l.* to 20*l.* each; work them three or four years, then fatten, or sell them to the graziers: in the latter case, for one-third more than the price at which they were bought in. Their food, two parts straw and one part sainfoin hay cut into chaff; otherwise they eat very little straw: turnips sliced into chaff, and at times some bran; corn never given, unless turnips and hay are wanting. Mr. Pinnal thinks that they cannot be in too high order for work, and that the reason why the use of oxen has in many cases failed, has been nothing more than bad feeding. They lie out in the yard in winter. All the farmers in this vicinity have more oxen than horses; and far more than they had, ten, fifteen, or

twenty

twenty years ago. He is very certain, that five oxen will do as much work as five horses; yet two oxen do not cost more than one horse.

Mr. Pinnall agrees with Mr. Tuckwell, that driving attentively is a great point in rendering oxen successful.

Mr. Tuckwell, of East Leach, has five or six teams of oxen, all Herefords. Mr. Faulkner, of Burford, two teams; I saw one, and the beasts very fine and large. Mr. Hemming, of Fullbrook, one. At Chilton there are three farmers that have them. Messrs. Bagnal and Pinnal, of Westall and Hollwell, have fifty to sixty at work—in a word, every body has some. Mr. Large, at Bradwell, has two or three teams; Mr. Gardner, three or four; Mr. Townsend, one; Mr. Tuckwell, of East Leach, three or four.

Mr. Secker, of Witney, works them successfully, and thinks that four in a plough will do as much work as four horses: Mr. Coburn is of the same opinion; but upon entering more particularly into the question of the comparison, I did not find that either of them had fully made up their minds upon this point.

Mr. Edmonds has worked oxen above fifteen years, and he is perfectly convinced that the advantage is greater than that of horses; they plough as well, and being bought at 20*l.* at three years old, they will on an average, after three years work, be worth from 26*l.* to 28*l.* if they have been fed as they ought to be.

Mr. Salmon finds the cross of Yorkshire cows by a long-horned bull, the best for working.

Another farmer in this vicinity keeps Herefords: breeds twenty in a year, and the beasts which he works (eighteen or twenty) are mostly bred by himself: three regular teams and one to spare in each: uses four or
five

five in a plough ; and on an average do as much work as horses : feeds on straw in winter, but does not work them then. An ox in very good condition, put to barley straw, will do well on it, and retain his flesh ; but if put poor to it, will do nothing. Is convinced that they are generally kept in too low condition.

" In many parishes having open fields, there may not be sufficient pastures for oxen, but in all enclosed parishes, where there are pastures, oxen might be introduced with great advantage, particularly in the vale, where there are no flints nor stones. There are teams of oxen in the county, chiefly belonging to gentlemen, which draw by collars and traces ; but the difficulty of procuring proper persons, that have been used to go with them, may be the reason of their not being in more general use. It is a custom with many farmers who do not breed their own colts, to purchase them for their teams when rising two years old ; some of which are sold for carriages, road waggons, or London drays, according to their strength and size, at four or five years old, and often yield good profit. But those horses which are kept after that time to be worn out, may be considered as an annual loss ; which is not the case with oxen, particularly as a profit will always attend an ox in case he meets with an accident, which is not so with a horse*."

" Oxen may do very well for a farmer that has a good deal of land to cultivate, and pasture to fat them upon at the proper age. In the north part of of Oxfordshire, oxen have been often tried, and as often discontinued."—*J. Chamberlin.*

One of the most interesting circumstances I met with

---

* Original Report.

in

in Oxfordshire, was the increasing attention paid to oxen, as beasts of labour. While the use of them has been going out in so many parts of the kingdom, and in so many others almost forgotten that such a practice ever existed, to find the use of them well understood, and every day increasing, was a spectacle which could not fail of giving me great pleasure. The speculative writers who labour hard to persuade their readers that this use of oxen is unprofitable, would do well to go to Burford, where they will find plain practical farmers, whose business is upon the largest scale, giving the preference to oxen, and yet well satisfied of the real use of horses. The minutes in that quarter are particularly satisfactory; and it will demand more ingenuity than has hitherto been exerted, to render them indecisive upon this question: in fact, the propriety of the use of these beasts, seems there to be so rationally established, as to bid defiance to whatever objections speculative ingenuity, or erroneous practice, can muster against it.

The result of those inquiries forms an object particularly interesting at this period. The annual import at present, of a million sterling in corn, no inconsiderable portion of which is for oats, and the well founded apprehensions that a deficiency may happen beyond the power of import to supply, united with the increasing population of the kingdom, prove that either the consumption of corn must be lessened, or the culture of it increased. That sufficient encouragement does not exist for cultivating those wastes, which, if cultivated, would be under corn, their present state is a lamentable proof: and as every proposition for a general enclosure has been successfully opposed, it does not appear probable that the quantity of corn produced will be

materially

materially increased: and when it is farther considered, that the prodigious rise in all the farmer's expense of tillage must have a constant tendency to convert arable land to grass, and that the general custom of refusing leases must gradually sap the foundations of the best husbandry in the kingdom; when all these circumstances are duly taken into the account, we shall hardly fail to admit, that the increase of the use of oxen is a point of material importance in the prosperity of the nation: and I may farther take the liberty of remarking, that the discovery of a district under the circumstances of that of Burford, at a period when the general prejudice is working against the use of oxen, would alone be sufficient to prove the great utility of such Reports as the Board of Agriculture have for so many years patronised. Oxen have been defended by reasoning as speculative as that wherewith they have been attacked: but so far has it been from producing cases of common farmers in the successful practice of a large business, in ten years increasing oxen as ten to one, that the opposers of the system have not used any arguments so weighty as those of urging the decline of their use in every part of the kingdom. This reasoning (in a measure fair, while proposed as reasoning only) has not been answered by an appeal to facts. The instance of His Majesty, and many great lords and gentlemen, have been put by, as inapplicable to a question of practice: but the case of Burford is of a different description Plain common farmers paying rent for their land, and wholly employed in the successful cultivation of it, greatly increase in their ox-teams; and proportionally lessening the number of their horses, is an instance free from every similar objection; and those, who wishing for further information, shall make inquiries of the persons

sons alluded to, will find that their practice does not originate in a want of understanding, but will be agreeably convinced, that they are kee., sensible, and intelligent. Reports that enable the Board to bring such facts to light, may continue to be condemned by some. It is conceived that others will not be wanting, who will draw very different conclusions

SECT. III.——SHEEP.

Mr. Fane has been for some years breeding gradually a flock of Spanish sheep. He began with a ram with which His Majesty favoured him, crossing some Ryland and South Down ewes; he afterwards purchased more in the early period of the King's sales, and has gradually got a considerable flock. It was with much pleasure I examined them; the wool is excellent, and the frame of many very good: they are regularly improving by attention to breeding, and by the progress of crossing the quarter, half, three-quarters bred ewes by rams of the whole blood, so that in a few years there will be scarcely any but Spanish blood in the flock. The wethers he kills come to 16 lb. a quarter. Mr. Fane is upon the whole very well satisfied of the profits of this breed; as he finds them hardy as any, after the first two or three days of their life. They fatten quickly, and are very healthy, and good milkers; upon his dry soil he is never troubled with the foot-rot: they give on an average $4\frac{1}{2}$ lb. of wool, at 4s. per lb.

The Earl of Macclesfield has a flock of South Downs, which I viewed with pleasure; they are very good ones.

Mr.

Mr. Lowndes, at Brightwell, has eighteen score of the same breed, and an Ellman ram.

Mr. Simmons, of Stokenchurch, kept the Berkshire sheep and South Downs in the same flock, and four toothed wethers, and both wintered and summered: many of the South Downs are fat, and sold to the butcher, but not one of the Berkshires.

Messrs. Percey, Sarney, Hayward, Goodchild, and Greenwood, in discourse on sheep at Mr. Fane's, all agreed that the South Downs were much superior in profit to both Berks and Wiltshires, and yet admitted a certain measure of merit to both those breeds: both are wearing out in favour of the other breed.

Sir C. Willoughby was for many years in the Berkshire breed of sheep, but changed them for the cross of South Downs, half Berks, half South Downs, which he has now had for five years, and finds them profitable. Last year Berks wool was 40s. pen tod, the half breed 50s. : this year Berks 38s. half breed 48s. : the former, seven or eight to a tod, the latter nine or ten. The objection to the Berks is, that they are too long in fattening, and are too large. He folds every thing except fat sheep : tried the experiment of leaving it off, and the sheep were sure to fold themselves, either in a long line upon the gravel road, or in a bunch knot under the trees.

His flock, 140 ewes,
100 wethers,
150 lambs,
50 to 60 to fat.

———

445

———

his farm near 400 acres.

It

It seems to be admitted in the vicinity of Baldon, that a Berkshire ewe with a New Leicester ram, forms a breed that is profitable: but upon the whole, Sir Christopher Willoughby prefers South Downs.

Mr. Davy, of Dorchester, was in the Berkshire breed, and had a good opinion of them; but he has changed them for South Downs, which he finds more profitable: he has 250 of them. In fatting a lot of South Downs with some Berks among them, they will be all gone before any of the others are ready for the butcher. Of crosses, that which he has the best opinion of, is half Leicester and half Berks, which is much better than taking any of the Wiltshire blood; the grand fault of the Berkshires, is taking so much time to make them fat.

Mr. Davy feeds his sheep with the straw of pease and beans; preferring the latter, but given plentifully, that they may only pick off the pods: they have turnips also.

In the Dorchester district, natural grass being extremely scarce, straw is given in large quantities to sheep. As soon as frosty mornings come, barley straw is given in the yards or standing pens; and afterwards bean and pea straw, which they are very fond of: they pick off the pods and tops, and do well upon this food. Bean and pea straw are sometimes carted to the field for their supply, and what they do not eat brought home to the yards: and this practice is pursued to the saving of many, many tons of hay: the dung thus made is found to be very good.

Mr. Thomas Latham, of Clifton, applies pease in fattening lambs, the mothers of which are at turnips. In six or eight weeks after falling, gives the pease in troughs. By the lambs running through

holes

holes in the hurdles, it is some time before they will take to them, but they come on gradually, till a score will eat a peck a day. Mr. Latham gives pease in this manner, to the price of 6*s.* a bushel, and finds that it answers.

Mrs. Latham, of Clifton, has one of the completest sheep yards, if not the most so, in this county : a shed surrounds three sides of it, in which are racks and mangers ; it is 31 yards long and 16 broad : the sheds five broad, and it does very well for 200 ewes. They are brought into it from four to six weeks before lamb-ing, and continued till it is over, going out in the day time. It is considered as a very excellent method, but attention should be paid that the dung does not accumulate, which by fermentation injures the sheep ; it is therefore carted out at several times.

Mr. Thomas Latham, of Clifton, approves much of the Berkshire ewe, covered by a Leicester ram, for selling fat lambs, and then fatting the ewe off : has long been in the Berkshire breed, which is a very good one, by way of a regular breeding flock. He sold the lambs at about a year old, before shearing. This Berkshire breed stands folding remarkably well, and as the farms at Clifton are without sub-division fences, to fold the sheep is a matter of necessity ; not that this is the only motive for folding, for Mr. Latham considers the practice essential in respect of manuring and tread-ing. The Berkshire sheep are strong, active, and able to travel, and fold uncommonly well ; with these good qualities, they have some bad ones, particularly in being long in fatting ; and it is principally with this view that he has taken the cross of Leicester blood : it is a system that is gradually becoming general.

Mr. Cripps, of Burcot, crossed his ewes last year with

with Leicester rams : there was a Leicester ram and a Berkshire, and the lambs got by the Leicester were fit for the butcher, long before those of the whole breed were in any order at all.

Folding goes on in summer for turnips, then for wheat, previous to sowing, and for a month or six weeks after Michaelmas, if the weather permits, and will be folding after the wheat is quite green. As soon as the wheat is sown, he drives his sheep two or three times up and down every land, and does not find that it hurts the sheep much. And this system of treading is of such importance, that he would not be precluded from it for 20s. an acre. The practice is considered as so harmless, relative to the sheep, that it is not uncommon for a flock-master to lend his flock for this purpose to a neighbour who has none. The object would not be answered by any system of common rolling. Last year he had a piece of clover land wheat, much eaten by slugs; and in November got broad-wheel carts loaded with stones, and drew them three horses a-breast, across, till the surface was well whelled upon. This operation killed the slugs, and the wheat proved as fine a crop as could be seen.

Mr. James Welch, of Culham, keeps the cross of Berks : ewes covered by a ram half Gloucester and half Leicester. The whole bred Berks is so fast going out that he could not sell them. He pens constantly; first for barley, then for turnips, and after that for wheat. Seventeen score in a pen of 75 hurdles, each six feet long—one night in a place. Such a fold is, as near as may be, a rood of land, or 34 square feet to each sheep.

At Bensington, South Downs, &c. Mr. Bonner's expression was, *the Berkshires are gone out of doors.*
Wiltshire

Wiltshire ewes are covered with a Leicester ram and sold in couples; Dorset, Berks, and South Downs, with the Leicester ram; but of these crosses, the Wiltshire beat; they are the first and heaviest, in his opinion. Next are the Downs, which are a good breed, but the Berkshire ewes are tender.

Mr. Shrubb, of Bensington, finds that the cross between the New Leicester and the Berks stands the fold well; and that if they are not quite so good for that purpose as the Berks, yet the superior profit more than makes amends: he is getting into the South Down gradually.

Mr. Newton, of Crowmarsh, used to keep the Wilts and Berks, but is now changing them for South Downs, which he is convinced is a more profitable breed. He finds that he can pen them as well as any other breed; nor will he admit the objection which has been made, of their lying irregularly in the fold: this has not occurred with him. Though he keeps above 1200 sheep besides lambs, he does not like a flock of more than 300, but divides, and keeps several shepherds. In order to meet the difficult season after turnips are gone, he sows rye for his flock, and trefoil is sown amongst the wheat, to be succeeded by turnips: this is spring fed, and it is beneficial to the turnip crop.

The Bishop of Durham buys in 300 South Down wether lambs at 27s. each; and has 150 three year old sheep besides. They go off fat at Christmas, &c. and the lambs are sold fat by April following, from turnips. Many vetches are eaten by the flock, and the farm horses fed on them.

Mr. Freeman, at Fawley court, has a various flock of different breeds, which he is going to change for 200 New Leicester ewes, and to keep round what is

bred

bred by them. Messrs. Foster and Hairbottle are decided enemies to the practice of folding; for though the benefit in corn is admitted, yet they contend that more is lost in the flock than gained in the crops.

Mr. Dean, of English, keeps Berks ewes, and crosses them by a ram whose breed is half Leicester and half Gloucester. He observes that the Berks are out of fashion, though the very best of sheep for folding: in this respect the South Downs are not to be compared to them; he has whole bred South Downs; his flock consists of seven score breeding ewes; he fats the lambs, giving them as much corn as they will eat; selling from Easter to Midsummer; he regularly winters 200 dry sheep, and as many more as his crops will permit.

The South Down breed are coming in at Maple Durham, &c.

Mr. Kelsey, at Whitchurch, buys 200 South Down lambs every year, which he folds and sells again, stores. He gives 24s. a-head for them, and sells before clipping, at from 29s. to 34s.

Mr. Cozins, of Golder, keeps the whole Berkshire breed: he has not yet got into the South Downs, because he must give as much as for Berkshires, that weigh six or eight pound a quarter more. He has some half Leicester and half Gloucester; also Wiltshires ram'd by South Down or New Leicester rams: these are good, and the lambs go soon to market. In the spring he gives 36s. for four-toothed Berkshire wethers: he folds them through the summer, and sells again at about the same price he gave, getting only the fold for his profit.

In the rich district of Thame there are many Wiltshire and Berkshire sheep kept, though not so common as they once were. Mr. Plaskett, of North Weston, being

being well convinced of the utter unprofitableness of those breeds, changed them for New Leicesters four years ago : he has shewn at Smithfield three times, and gained prizes twice. His neighbours do not yet come to him for tups; yet it is certain, that he keeps five instead of three of theirs. His farm contains 130 acres, and last year he wintered 300, including lambs, that is,

Ewes, ... 110
Lambs, ... 120
Shear hogs and two shear, ... 70
____
300

Besides these sheep, he had eight large short-horned cows, and 40 acres arable, of which sixteen were tur- nips, Swedes, and rape ; an account which sufficiently proves not only the richness of the land, but the good- ness of the breed.

Mr. Creek, of Aston, has been for many years in the New Leicester sheep, and with great success : he has hired many tups of Mr. Stubbins, and is well persuaded that for the stonebrash of Oxfordshire, there is no breed that comes near them. He and Mr. Wing and Mr. Filmer, positively assert that they produce a lamb for every ewe on an average of years, and the number kept is certainly a proof that the system here is profitable : it amounts to a sheep per acre on some farms ; and on others in the proportion of three to four, or 300 sheep on 400 acres ; not a regular breeding flock of selling lambs (though some do that, and at 35s. a-head, in August), but keeping round : folding these sheep is universal, but never to drive them far ; a fold of twelve dozen hurdles for 100. The hurdles are 6 feet long, 24 yards square, which make 576 square yards per sheep, one night in a place; is very thin folding indeed : it is 850 sheep folding an acre per night. Mr. Creek contends that folding is necessary to prevent their being too fat : it is curious to see how facts elicit in conversation, explanatory of collateral inquiries : if the inquiry is into the merits of folding, we often hear that it is not prejudicial to a flock, and that you can keep as many folded as without folding : but here we have the important confession, that folding keeps sheep from being fat. Why then is not the conclusion in common sense, that if you did not fold you could keep more ? Mr. Wing folds them in a littered yard, to lamb in.

Lady Gardner has 50 breeding ewes, of much more Leicester than Gloucester blood ; has had a ram every year from Mr. Creek: 70 lambs, one year with another. Sells off ten or twelve double couples at 3*l*. 3*s*. to 3*l*. 13*s*. 6*d*. in April; wether lambs go fat to the butcher in June or July, at 28*s*. to 30*s*.; the fat culled ewes 20 lb. a quarter : clip five to five and a half to a tod. In the winter they have hay, and the lawn. Lady Gardner manages her beautiful lawn extremely well : the grass, exclusive of wood and water, is under 100 acres; she has very little arable, and no turnips, but keeps 134 sheep, 16 cows, and other cattle, and nine horses of all sorts. She is fond of the place which her taste decorates ; her green-house, her garden, her flock.

Mr. Forster, of Bignal, is in the cross between the New Leicester and Cotswold breed ; though his farm is on the stonebrash, he prefers them to any others. I saw four tups on his farm, a lot of ewes, and another of theaves : they are good sheep. Objecting to their not breeding well, his nephew assured me that they did not find it so, and that 100 ewes would always bring

above

above 100 lambs. His theaves shew more *blood* than the ewes, and the ewes full as much, and some more, than the tups; but if he goes on and breeds for delicate heads with blueish countenances, he may find what so many others have found: at present, and with the rams I saw, he is not in danger, and the profit of such sheep is unquestionable.

The Duke of Marlborough has a good flock of half Leicester and half Cotswold; folds only on wheat; some South Downs and some Welsh. Mr. Palmer thinks that while the flock was kept very much in the Leicester blood (that is, much more than at present) he had more barren ewes, nor were the lambs so strong or healthy; more and better since he took more of the Gloucester blood; the South Downs he thinks a good hardy breed: the Welsh are merely for the Duke's table.

Mr. Palmer has supplied of all sorts of meat, to the amount of one ton per week to the house, 320 lb. per diem.

Mr. Sotham considers half Leicester and half Gloucester as the best breed; are of a good size; yield plenty of wool, and bear penning well: and he thinks no size so profitable as from 22 lb. to 24 lb. per quarter, for shear hogs to be fatted by old Christmas. They are in this country much subject to the wood-evil, which attacks their joints, backs, and legs: it is attributed to the land. Some lambs also are affected by it; woods and furze are likewise supposed to occasion it; those bred upon the farm escape best. It may generally be estimated, that every acre carries a sheep and a lamb: perhaps on an average of the county, 300 are sheared on a farm of 400 acres. Mr. Sotham does not pen in summer: he wants water, and therefore the sheep are left out for the dews. He is clear that more

sheep

may be kept without penning than with it; and more is lost in the sheep than the land gains. In summer they lie under hedges, and if penned, must go to the pens empty: they do far better without it.

Mr. Pratt, of Fowley, has a very high opinion of folding sheep, though all in the country are long woolled; and he is confident that more sheep may be kept on a farm by folding than without it: his reason is, that they stain their food so much less, by taking them from their feeding ground, where they are sure to rest too much in the same place, if not driven from it. *Mr. Pratt, do you ever fold fatting sheep?* Never. *Why not?* They will not do so well if they are folded. *Mr. Pratt, can that be beneficial to an ewe which keeps a wether from fattening?*

Mr. Bowles, of North Aston, is chiefly in the Leicester breed, from Mr. Creek, his neighbour; and the specimens I saw, appeared to be bred with a good deal of attention: many of them are very handsome, and from the few particulars I could gain in the absence of their owner, I doubt not, very profitable.

At Great Tew, a flock of 60 breeding Gloucester ewes, 70 lambs, 70 theaves and wether tegs; 70 kept or sold fat.

Mr. John Hewett, of Goring, fattens many on oil-cake: he does not conceive that the improvement of the sheep quite pays for the cake, unless laid in at a reasonable rate, but the advantage by the manure is very great, and beats every other: the cake is given in troughs in a pen, moved every day; usually on wheat stubbles for oats. Last autumn, 1806, cakes were 14*l.* 14*s.* per 1000, but rose afterwards to 18*l.* 18*s.* There is a good deal of cake feeding sheep and beasts in that neighbourhood: the cakes from Twickenham mills.

Mr.

Mr. Wyat, of Water Eaton, is in the Leicester breed: he hires tups of Mr. Buckley and others, and by this means breeds tups to lett to those who cross the Cotswold breed. I saw his flock, and found that he was breeding with attention and success: he has some ewes that are as fat as he can wish them to be. His neighbour Mr. Rowland breeds none, but buys in for fatting.

Mr. Weyland, of Wood Eaton, keeps the cross of half Leicester and half Berkshire, and gets rather more than a lamb to every ewe. Last year he reared 110 lambs from 100 ewes: he pens them the year round; but they are driven only a very short distance.

Measured the sheep pen on newly sown wheat of Mr. Wing, at Aston; it contained 704 square yards for 171 sheep, or $4\frac{1}{2}$ square yards per sheep.

The Rev. Mr. Filmer, whose stacks, stubbles, tillage, and many other circumstances, shew much and successful attention, is decidedly of opinion, that the long woolled sheep are very profitable on the stonebrash land.

The economy of a stonebrash flock may be gathered from the following circumstances of that of the Rev. Mr. Filmer, at Hayford.

In 1807, sheared 287.

150 breeding-ewes,
70 shear hogs, 24 lb. per quarter,
67 theaves,
___
287
130 lambs,
___
417 total at shearing.

In

In September following, the number,

Breeding ewes, ... 150
Barren, sold fat, ... 27
Culled, ... 20
                                                      —— 47
                                                      ————
                                                          103
Added theaves, ... 53
                                                      ————
Number now put to ram, ... 156
Lambs, ... 130
Shear hogs for turnips, ... 70
                                                      ————
Number in September, ... 356
                                                      ————

Sale.

27 barren ewes fat, at 48s. ....£.64 16 0
70 shear hogs, at 63s. 220 10 0
20 culls, at 48s. 48 0 0
287 fleeces, at 6lb. 90 0 0
                                                      —————————
                                            £.423 6 0
                                                      —————————

Expenses of the flock—50 acres of turnips, 70 of grasses, 50 in the common course, and 20 of the second year. The shepherd 12s. a week, the year round, with help at lambing time, and turnip feeding; shearing, 5s. per score; washing and various expenses, 20s. per annum.

The breed preferred in the country round Chipping Norton, is the half cross of Leicester and Cotswold. Mr. Kimber, of Little Tew, does not at all approve of South Downs, being well convinced that the profit of them is not nearly equal to that of a larger sheep. He admits that the blueish countenance and delicate heads

of

of high bred Leicesters, is a sign of little fertility in
the production of lambs, but the Cotswold single cross
remedies it entirely.

Mr. Huckvale, near Chipping Norton, letts many
Leicester rams. The farmers around that place are
much in the Leicester breed, but many prefer a cross
of the Cotswold.

Mr. Kench, on 1000 acres of stonebrash, keeps 300
breeding ewes, half Cotswold and half New Leicester;
his flock is consequently (as he keeps round) 1200.

> 300 ewes,
>
> 300 lambs, { 150 ewes,
>   { 150 wethers,
>
> 300 hogs and theaves,
>
> 300 shearlings going off; varied by culls,

fatted; theaves set for stock, and rams.

Mr. Salmon, of Hardwick. has New Leicesters: to
332 breeding ewes last year, he had 380 lambs, and
generally has more lambs than ewes; he pens them,
and thinks it necessary.

Mr. T. Wyat, of Hanwell, has been long and deeply
in the breed of the New Leicester sheep, having gone
to high prices for tups; such as 200 guineas for the
ride of one, and himself letts at 40 guineas: I examined
some of his tups and lambs, &c. and found them such
as did credit to his judgment; he has the best tup I
have yet seen in Oxfordshire.

Mr. T. Payne, of Drayton, finds the New Leicester
by far the most profitable stock. He keeps 300 of them
(lambs included in the system of keeping round and
selling fat) on his farm of 200 acres; and has sold 21
shearling hogs for 100 guineas, fed without corn, and
having only Swedes and cabbages. He has two standing folds for them to lamb in, with sheds, and racks,

and

and mangers, along one side, and hay stacked close for
convenience of feeding: well executed, and does him
credit.

Mr. Payne, and Mr. Wyat, of Hanwell, have a house
on four wheels, to cover a man and boy in using a machine for slicing Swedes; a man and boy will cut
enough for 100 sheep, and give them in troughs: no
scooping or waste, and answers better than any thing;
and also gives under sheds, and they thrive much
better, most advantageously, never go backward, but
thrive well in all weather.

In feeding a field of Swedes in winter, builds a temporary shed of thatched hurdles, under which they go
at pleasure; at one end a repository of the roots to
slice, and give in troughs varied over the land; the
hurdles being moved on, and the smaller roots left for
them to eat; sheep thus do much better. A little litter
scattered about the pen, &c. and the dung afterwards
thrown up, and spread where most wanted: trampling
avoided. Has found it an excellent method.

At Atterbury they were in the Warwick breed of
sheep (long woolled), by which I suppose is meant
something like the Old Leicester; now they have got
very much into a cross of the New Leicester, which
are found far more profitable: they keep breeding
flocks, and fold them. Mr. Bellow folds his ewes and
theaves, but not double couples, or fatting sheep: the
New Leicesters stand it well, but they are never driven
far. His flock does not lie out of fold three months in
the year.

Mr. Davis, of Bloxham, has a mixture of Leicester
and Gloucester; on the enclosure he bought a flock
and bred from them; the Leicesters grafted on the
Gloucesters are the best: does not fold, but in summer,

mer, on large tracts, it may be beneficial; generally more lambs than ewes: South Downs better on lighter lands.

Mr. Turner, of Burford, keeps half Leicester and half Gloucester; pens them, and keeps one sheep and lamb per acre, over his whole farm. He is not an advocate for penning except in eating vetches and turnips. He has no doubt of more sheep being kept upon a farm without folding than with it. Shear hogs 30 lb. a quarter; his tegs three and a half to a tod; ewes four and a half.

Mr. Pinnal, of Westall, is in the same breed, a cross which he thinks necessary, on the side of Gloucester, for size and wool, and on that of Leicester for inclination to fatten: feeds his flock in April, on Swedish turnips and hay; never folds them: in all the heat of summer, folding would starve them. At that season they do not feed in the day time. He admits that the dung is not dropped regularly, and that it would be disposed more advantageously by folding; but this is a small object compared with the benefit of the sheep: they bring in the proportion of 250 lambs to 200 ewes: tod four; New Leicester five to six; and all Cotswold three and a half.

Mr. Pinnal's shearlings, killed a little before they would be two shears, come to 27 lb. or 28 lb. per quarter.

Mr. Secker and Mr. Coburn, of Witney, have a mixture of the Leicester, but do not take it often, as that breed lessens the wool and gives a smaller carcass, but brings them to fatten at a younger age than if all Gloucester blood. Shear hogs weigh 22 lb. to 24 lb. a quarter, and clip five to a tod: folds a little on the open

open field, but more to keep them from straying than from any approbation of the system.

Mr. Edmonds has bred with attention for fifteen years past, with a small cross of the Cotswold well selected, and coming near to New Leicester in the hands of Mr. Haines, Mr. Kimmer, Mr. Freeman, Mr. Creek, and from Mr. Astley of Leicestershire. His object has been, and is, to retain the weight and length of wool as near as possible of the Cotswold, with the disposition to fatten that can be gained by means of Leicesters.

Native Cotswolds, if to be found, would be at two shear from 28 lb. to 32 lb. a quarter : a long sheep, not full in the sides, sharp in chine, not full in fore flank, coarse in the bone, not strait but good in the hind quarter ; will not fatten so early as when crossed ; and of wool, two shear wether three and a half to the tod ; the New Leicester is calculated to correct every one of these deficiencies, and brings a greater disposition to fatten. Between all Cotswold and all Leicester, average difference of wool, 3 lb. : has not suffered at all in the number of lambs, but has been careful in his selection, as he has seen individuals which he should not choose to depend on for getting lambs. Mr. Edmonds never folds, and of course does not approve of it. Thinks upon all hill farms, that stock is the grand object, or ought to be, and therefore if the sheep are folded, that rule is broken, and corn made the first object : on an average, has as many lambs as ewes.

It is remarkable, that in this parish (Rousham), no sheep have ever been known to rot, even whilst it was unenclosed, though the disease has frequently been very fatal in the adjoining ones.

There

There are two circumstances explained in the preceding minutes, which are to the credit of the sheep husbandry of Oxfordshire: the gradual introduction of South Downs to the exclusion of the Berkshires, and the general number of sheep kept in proportion to the extent of the farms, which through various parts of the county is a point of considerable merit.

" A ground in this neighbourhood, part arable part pasture, the arable on a declivity; after much rain, the washing from the arable on the pasture, often occasioned the rot : the whole has been laid down in grass seven years, and not a sheep has taken the rot since. —*Thomas Curtis, Nether Worton, Deddington.*

SECT. IV.—HOGS.

" The *pigs* in most esteem with farmers, are those which will prove of a large size when fat; but I am convinced no sort is so profitable to a small family, or a poor man, as the Chinese, or a cross between that kind and the breed of the country ; because they are maintained and fatted on less food than others, and by a cross they will come to great size.

" It is worthy observation, that many boars are fed for the purpose of making brawn, which forms a conderable article of trade at Oxford, and other parts of the county*."

Mr. Filmer's hogs are of a very good Berkshire sort ; short noses, broad backs, and smooth hides.

---

* Original Report.

Observing

Observing in Mr. Kelsey's farm-yard at Whitchurch, many Berkshire pigs, I found he kept four breeding sows, fed in winter on pollard and raw potatoes; and in summer on vetches or clover. He does not think they form an article of profit except for the dung, which amounts to a considerable improvement through the whole of the farm-yard.

The Berkshire breed is the most common throughout the county.

SECT. V.—RABBITS.

Mr. Fane has raised a small building for keeping rabbits in hutches, for the sake of the manure; he has some hundreds, and meditates a second building for doubling the number; they make a load of manure per week, and as two manure an acre, they are the means of well dressing 26 acres annually: he does not conceive that they yield any other profit, nor is it necessary; for to be able to sell so much food at home, and pay attendance, the profit of manuring such a breadth of land is considerable. He sends three dozen a week to London.

CHAP.

CHAP. XIV.

RURAL ECONOMY.

——◆——

SECT. I.—LABOUR.

AT Stokenchurch, thrashing wheat, 4*s.* a quarter; barley, 1*s.* 8*d.*; oats, 1*s.*; reaping wheat, 9*s.* to 14*s.*; mowing barley, 2*s.*; mowing grass, 2*s.* 6*d.*; women per day in hay, 10*d.* and beer; in harvest, 1*s.* and board; head ploughman or carter, 11*l.* 11*s.*

At Tetsworth :

Labour per week in winter, with beer, ... £.0	9	0
———— spring, ...	0	10	0
———— hay, from 2*s.* per day to	0	3	0
Reaping wheat, from 9*s.* per acre to	0	12	0
———— beans, from 5*s.* 6*d.* to	0	7	0
Mowing barley and oats, 2*s.* 9*d.* to	0	3	6
Women in hay, per day, with beer, ...	0	1	0
———— at other times, ...	0	0	8
Head man's wages, 10*l.* 10*s.* to	12	12	0
Thrashing wheat per quarter, ...	0	3	6

Quartern loaf, 9*d.*; butter, 1*s.* 3*d.*; beef, 7½*d.*; mutton, 6¼*d.*; hay, 2*s.* 6*d.* per truss.

At Wormsley :

Labour in winter, ... £.0	10	0
———— spring, ... 0	11	0
———— hay, with beer, ... 0	11	0
Reaping wheat, from 9*s.* to 0	12	0

Mowing

	£	s.	d.
Mowing barley, raking, from 3s. to	0	3	6
———— grass, with beer, ...	0	2	6
Women in hay, with beer, ...	0	1	0
————————— harvest, 1s. to	0	1	3
————————— the rest of the year, ...	0	0	8
Thrashing wheat, per quarter, ...	o	4	0
———————— barley, 1s. 10d. to	0	2	0
———————— oats, ...	0	1	4

Ploughing, three horses, one acre a day.

At Baldon:

	£	s.	d.
Reaping wheat, per acre, ...	0	10	0
———————— beans, ...	0	6	0
Mowing barley, ...	0	1	6
———————— oats, ...	0	1	6
Labour by the week, winter, ...	0	9	0
In hay mowing (no beer), ...	0	14	0
Mowing grass, per acre, beer, ...	0	2	0
Carter (with beer), per week, ...	0	10	6
Dibbling beans, per acre, ...	0	4	6
Women, ...	0	0	8
Ditto in harvest, 1s. to	0	1	6

All the women about Baldon reap in harvest, and their common earnings 3s. a-day. Too much cannot be said in commendation of this industry. A man and his wife will reap three-fourths of an acre in a day. It is generally bagg'd ("fagg" here) for cutting low: where fuel is scarce the family that reaps has the stubble. This is bad, for it is in fact burning their dung-hills.

At Bensington, hoeing turnips twice 6s.; reaping wheat 9s., and 12s. have been given; beans 5s.; hoeing beans 5s. each time. By the day in winter 1s. 6d. but if the men have no piece-work, 10s. a-week. In

harvest

harvest 10*s.* and board; if no board 20*s.*; women in hay 8*d.* and beer.

At Clifton: thrashing wheat, 2*s.* 9*d.* per quarter; barley, 1*s.* 6*d.*; oats, 1*s.* 3*d.*; beans, 1*s.*; pease, 1*s.* 4*d.* Reaping wheat, 10*s.* per acre; beans, 6*s.* ditto. Mowing barley, 2*s.* 4*d.* a-day, and beer; grass, 2*s.* 6*d.* per acre, 1 quart of beer a-day. Labour in winter, 9*s.* a-week. Women 8*d.* a-day; in harvest 1*s.* 2*d.* Dibbling beans, 4*s.* per acre; hoeing turnips, 5*s.*; if in harvest, 6*s.* per acre. Carter 10*s.* a-week and beer.

At Mungwell: a carter 11*s.* a-week the year round. Common labourers 10*s.* except in harvest, when 10*s.* 6*d.* and board. Reaping wheat 10*s.* per acre. Mowing and cocking barley, 3*s.* 6*d.* Meadow, 2*s.* 6*d.* and beer. Thrashing wheat, 3*s.* 6*d.* per quarter; barley, 1*s.* 6*d.* ditto; oats, 1*s.* 4*d.* ditto.

At Maple Durham: in winter, 1*s.* 8*d.* and beer; spring and hay, the same; in harvest, 10*s.* a-week, and board. Reaping wheat, 9*s.* to 13*s.* an acre; mowing, cocking, and raking barley and oats, 3*s.* 6*d.* Thrashing barley, 1*s.* 8*d.* per quarter; wheat, 3*s.* to 3*s.* 6*d.* and beer.

At Golder: labour in winter, 10*s.* a-week; in hay, 12*s.* one quart of beer a-day; mowing grass, 3*s.* per acre, statute measure; mowing barley, 3*s.* Reaping wheat, 10*s.* per acre; beans, 7*s.*; thrashing wheat, 3*s.* 3*d.* per quarter; barley, 1*s.* 8*d.*; beans, 1*s.* 6*d.*; hoeing turnips, 5*s.* per acre; dibbling beans, 5*s.* per acre; women, 7*d.* a day; in hay, 8*d.* and beer; in harvest 1*s.* and beer; when raking after the waggons, &c. dinner.

Near Oxford: labour in winter, without beer, 12*s.*; in spring, 15*s.*; in hay, 2*s.* 6*d.* per acre, and beer, and dinner when carting. Reaping wheat, 10*s.* per acre;

beans,

beans, 7s. ; mowing barley and oats, 2s. 6d. per acre ; women in harvest, 1s. a day, and beer ; in hay, 10d. Carter, 1s. a-week more than labourer, with perquisites. Horse, one bushel of oats a-week ; thrashing wheat, 4s. 6d. per quarter ; barley, 2s. 6d.

At Wood Eaton : in winter, 9s. per week, and no beer ; in spring, 10s. per week, and no beer ; in hay, 12s. per week, with beer. Mowing grass, 2s. 6d. to 3s. per acre, one gallon of beer to an acre ; women, in hay, 8d. a-day, and beer ; in harvest, 1s. and beer. Reaping wheat, 12s. an acre, sometimes more ; mowing barley, 2s. and 2s. 3d. with beer ; dibbling beans, 2s. per bushel ; hoeing turnips, 7s. per acre ; head man's wages, twelve guineas a year. Four horses plough three roods a-day, sometimes more.

At Wendlebury : thrashing wheat, 3s. 6d. and 4s. per quarter ; thrashing barley, 1s. 10d. and 2s. ditto ; thrashing oats, 1s. 8d. Labour in winter, 9s. and beer ; in spring 10s. 6d. Reaping wheat, 8s. with beer ; mowing barley and oats, 2s. 6d per acre, and no beer ; mowing grass, 3s. and one bushel malt ; head man, 14l. to 16l. per year. Four horses plough fallow ; three other land. Hoeing turnips, 7s. 6d. and one quart of ale a-day.

At Bignal and Caversfield : labour in winter, per week, 9s. ; in spring, 10s. ; in summer, 12s. ; in harvest, 1l. Reaping wheat, from 7s. 6d. to 10s. ; mowing barley, 2s. per acre ; grass, 2s. 6d. ditto ; oats, 1s. 9d. per acre ; thrashing wheat, 3s. 6d. and 4s. per quarter ; thrashing barley, 2s. 6d. ditto ; thrashing oats, 1s. 8d. ditto ; head man's wages, from twelve guineas to thirteen guineas. Dibbling beans, 2s. 6d. per bushel ; hoeing turnips, 8s. to 9s. per acre. Women, 6d. to 8d. in winter ; in harvest, 1s. 6d.

At

At Heyford: labour in winter, per week 9*s*. with small beer; in hay, 12*s*. and no beer; mowing grass, 2*s*. per acre; ray-grass, 2*s*. 6*d*. per acre. In harvest, one guinea a week, and beer; or, reaping by the acre, wheat, 10*s*.; mowing barley, 2*s*. 6*d*.; mowing oats, 2*s*. a-day, and beer. Women weeding, 7*d*.; in hay, 8*d*. a-day; in harvest, 1*s*. and beer. Thrashing wheat, 3*s*. 6*d*. per quarter; barley, 2*s*.; oats, 1*s*. 6*d*.; hoeing turnips, 8*s*. per acre.

At Bampton: thrashing wheat, 5*d*. to 7*d*. per bushel; barley, 1*s*. 6*d*. to 1*s*. 8*d*. per quarter; beans, 1*s*. 6*d*. ditto; reaping wheat, 5*s*. to 10*s*.; mowing barley, 1*s*. 3*d*.; mowing beans, 1*s*.; mow all and carry loose. Day work in winter, 1*s*. 6*d*.; in summer, 1*s*. 8*d*. and beer; in harvest, 15*s*. per week, and beer.

At Burford: thrashing wheat, 5*d*. to 6*d*. per bushel; barley, 1*s*. 6*d*. to 2*s*. per quarter; oats, 1*s*. 2*d*. to 1*s*. 6*d*.; reaping wheat, 8*s*. per acre, and beer; mowing spring corn, 1*s*. 6*d*. to 1*s*. 8*d*.; seeds and sainfoin, 1*s*. 4*d*. to 1*s*. 6*d*. some 2*s*.; meadow, 2*s*.; day work in winter, 8*s*. to 10*s*. a week, and no beer: 10*s*. per week lasts to harvest. Harvest, 15*s*. to 18*s*. a week, and beer. Women per day in hay, 8*d*. to 9*d*.; head man's wages, twelve to fourteen guineas.

Near Banbury: labour in winter, 10*s*. per week, and no beer; in spring, 11*s*.; in harvest, 12*s*. and board; mowing grass, 2*s*. 6*d*. per acre, and beer; women in hay, 9*d*. and beer; in harvest, 1*s*. and beer. Reaping wheat, from 10*s*. to 12*s*. per acre; mowing barley, 2*s*. per acre, and beer; thrashing wheat, 4*s*. per quarter; barley, 2*s*. 4*d*.; oats, 1*s*. 8*d*.; beans, 1*s*. 6*d*.

" The price of labour will always be regulated by demand: let long leases be given to tenants, and a commutation for tithe, and the *demand* for labourers will

keep up the prices of labour, so as to supesrede the necessity of any interference of the legislature, or the magistrates, for the purpose of increasing the price."
—*J. Boys.*

General Average of the preceding Minutes.

	£.		
Day labour in winter, ...	0	9	6
——————— spring and hay, ...	0	11	6
——————— harvest, ...	0	19	0
Women per diem in hay, ...	0	0	
————————————— harvest, ...	0	1	2
Reaping wheat, ...	0	10	0
Mowing barley, ...	0	2	4
Thrashing wheat per quarter, ...	0	3	7
——— barley, ...	0	1	10
——— oats, ...	0	1	6
——— beans, ...	0	1	5
Mowing grass, ...	0	2	8
Hoeing turnips, ...	0	6	6

SECT. II.—PROVISIONS.

At Wormsley: mutton, $7\frac{1}{2}d.$; beef, only cow's, 8*d.*; veal, 8*d.*; butter, 1*s.* to 1*s.* 6*d.*

At Baldon and Oxford: beef, $7\frac{1}{2}d.$; mutton, 7*d.*; bacon, 1*s.*; butter, 1*s.* 4*d.*; quartern loaf, 9*d.*; coals, 1*s.* 6*d.* per cwt.; Wednesbury ditto out of boats, 1*s.* 2*d.*; potatoes, 10*d.* to 1*s.* the peck; cheese, 9*d.*; billet wood, 1*s.* 2*d.* per cwt.

At Clifton: coals, 1*s.* 3*d.* per cwt.; bread, $8\frac{1}{2}d.$ quartern loaf; beef, 7*d.* per lb.; mutton, 7*d.*; bacon, 11*d.*; butter, 1*s.* 3*d.*

At

At Golder: bread, 9*d.* a quartern loaf; beef, 7½d. per lb; mutton, 7*d.*; bacon, 11*d.*; butter, 1*s.* 3*d.*

At Thame: coals, 2*s.* 2*d.* per cwt.; bread, 9½d. per quartern loaf; butter, 1*s.* 4*d.* per lb.; bacon, 1*s.*; beef, 8*d.*; mutton, 7½d. and 8*d.*

At Hayford: bread, 9*d.* per quartern loaf; bacon, 1*s.* per lb.; beef, 7½d.; mutton, 7*d.*; butter, 15*d.*; hay, 4*s.* per cwt.; coals, 1*s.* 1*d.* per cwt.; before the navigable canal, about twenty years ago, the people were greatly distressed for firing, wood being scarce; they were obliged to burn straw, &c. or any thing they could procure; but now as well supplied with coals as any village in Oxfordshire.

At Banbury: bread, 9*d.* per quartern loaf; butter, 1*s.* 3*d.* per lb.; beef, 8*d.*; mutton, 7*d.*; coals, 1*s.* 2*d.* per cwt.; wood, 1*s.* per cwt. Common cottage, 3*l.* a year, and taxes.

At Bicester and Wendlebury: beef, 7½d. per lb.; mutton, 7*d.*; pork, 8½d.; butter, 1*s.* 2*d.* In 1789, it was 7*d.* and 8*d.* in winter and summer; now, 1*s.* 3½d. in winter, and 1*s.* 1½d. in summer. Coals, 1*s.* 3*d.* per cwt. allow them to the poor, in winter, at 1*s.*

CHAP. XV.

POLITICAL ECONOMY.

SECT. I.—ROADS.

I REMEMBER the roads of Oxfordshire forty years ago, when they were in a condition formidable to the bones of all who travelled on wheels. The two great turnpikes which crossed the county by Witney and Chipping Norton, by Henley and Wycombe, were repaired in some places with stones as large as they could be brought from the quarry; and when broken, left so rough as to be calculated for dislocation rather than exercise. At that period the cross roads were impassable but with real danger. A noble change has taken place, but generally by turnpikes, which cross the county in every direction, so that when you are at one town, you have a turnpike road to every other town. This holds good with Oxford, Woodstock, Witney, Burford, Chipping Norton, Banbury, Bicester, Thame, Abingdon, Wallingford, Henley, Reading, &c. &c. and in every direction, and these lines necessarily intersect the county in almost every direction. The parish roads are greatly improved, but are still capable of much more. The turnpikes are very good, and where gravel is to be had, excellent.

SECT. II.——CANALS.

The Birmingham canal is of immense importance to Oxfordshire, immediately connecting London, through Oxford, with Birmingham, Manchester, and Liverpool, and with the Wednesbury collieries.

SECT. III.——MANUFACTURES.

Witney—Formerly noted for the weaving manufactory, now very much declined. Instead of upwards of 400 hands employed about five years ago, there are now (1807) about 150. Their earnings are from 10s. to 12s. per week : 12s. the utmost—average about 11s. Women earn from 6d. to 8d. per day ; average 7d. Boys from twelve to fifteen years old, 2s. 6d. to 3s. per week. Children employed in abundance, from six to ten years old : earnings, from 1s. to 1s. 6d. per week. Men women, and children, about 1000.

The manufacture at Witney declined for so many years, that there was very little expectation of its ever reviving. When I was there 39 years ago, there were above 500 weavers in the place, but it sunk gradually to about half that number, and even lower ; and very great distress was found in the place; and it was threatened with the utter loss of every means of giving bread to its numerous poor : but very fortunately for its inhabitants, the spinning jennies were introduced, with other machinery, especially the spring looms, by which one man does the work of two. As much wool (skin,

not

not fleece wool) is wrought here as there was 40 years ago, which was then estimated at 7000 packs, and trade is increasing ; machinery at present earns 4000*l*. a year, and the place, I was assured, is flourishing. But in respect to the state of the working hands, the medal must be reversed ; for the former state of the manufacture having nursed up a great population, the effect of the introduction of machinery gave with such a population the power of keeping down wages in such a manner as to deprive the poor of any share in, or at least leaving them a very small one in, that prosperity which has pervaded the kingdom, and so greatly raised the general wages of labour. In 1768 I found that 10*s*. to 12*s*. a week were the earnings of the weavers ; and they are now but 12*s*., with the employment of less than half the number that were here in the former period : women and children earn as above-mentioned.

In such a state of things, the masters and the fabric may flourish, but it cannot be contended that the labouring hands do the same. Such a plenty of hands as keep down the price of labour, must, with any degree of management, make a fabric flourish as well as it can flourish, relative to various other circumstances, such as the want of vicinity to coals, which benefit in the north has been so long attracting so many of the manufactures of the kingdom.

I heard of certain regulations at the Hall, long ago established, for keeping up the credit of the fabric, from which other places being exempt, have rivalled and injured Witney ; but such circumstances demand a much more minute examination than would be proper to detail in an agricultural work. The fact of the present poor-rates, which are 11*s*. in the pound, rack rent, is a strong confirmation of the preceding

parti-

particulars; and by the way, shews how very pernicious manufactures *may* become to the landed interest.

From the Rectory being worth 2000*l.* per annum,
Witney may be supposed to be an agricultural as well
as a manufacturing parish; and the land is thus burdened by a fabric, which for a long period declined,
and revives *not to the ease of the farmers.*

But do not the rise of rents make amends? By no
means. In 1768 the general enclosed rent was 20*s.*
and open fields 7*s.* to 11*s.* Enclosures have more
than doubled: field land has not doubled. Yet in 40
years the merely agricultural parts of the county have
in many parts quadrupled, and in all much more than
doubled. In the former period blankets were made to
3*l.* a pair; they are now made up to 5*l.* a pair. Provisions in 1768; mutton, 4*d.* to $4\frac{1}{2}d.$ per lb.; beef,
4*d.* to 5*d.* per lb.; veal, $3\frac{1}{2}d.$ per lb.; bacon, 8*d.* per
lb.: butter, 6*d.* to 7*d.*—At present (1807), mutton, 7*d.*
per lb.; beef, $6\frac{1}{2}d.$ to 7*d.* per lb.; veal, $7\frac{1}{2}d.$ per lb.;
bacon, 10*d.* per lb.; butter, 10*d.* to 1*s.* 1*d.* per lb.

This rise deserves attention, and should silence
much of the vague assertions heard in common conversation, of the enormous rise of provisions: the
contrary has been the case here: the rise of some is
great; but when we look back through this long period of accumulating national wealth and prosperity,
and compare this rise with a multitude of other circumstances, it will cease to surprise, and appear, as
in truth it is, moderate.

The

The rise is, Mutton, ... 64 per cent.
                  Beef, ... 50
                  Veal, ... 114
                  Bacon, ... 25
                                              ______
                                              4)253
                                              ______
                              Average meat, ... 63
                                              ______

Bacon, the lowest of these, is the article that most concerns the poor.

                  Butter, ... 76

The town of Thame seems to labour under a very depressing poverty : they appear to have no means of industry. A very few among them make a little lace, but there is nothing flourishing in the fabric; and the price of coals is greatly against the comforts of the poor. They are 2s. 2d. per cwt.

Woodstock.—About the beginning of the last century, the manufacture of polished steel was introduced here by a Mr. Medcalfe, which at one period was very flourishing, and employed a considerable number of hands; but from the cheapness with which articles of the same description, namely scissors, watch-chains, and a variety of trinkets, can be produced at Birmingham and Sheffield by means of machinery, a manufacture, however elegant, that is wholly completed by hand, was not likely to stand a competition, and consequently it does not now employ above ten or twelve persons, who probably may earn from a guinea to a guinea and a half each per week.

The Woodstock polished steel is wholly made of the old nails of horses' shoes, formed into small bars, and
                                                applied

applied according to the various purposes required. Its lustre is extreme, and it possesses the property of being repolished at a trifling expence to its original brilliancy, though covered with rust, an advantage it maintains over the articles which are generally substituted in its room. A chain made here, and weighing only two ounces, sold for 170*l*. A box in which the freedom of the borough was presented to Lord Viscount Cliefden, cost 30 guineas; and a garter star for His Grace the Duke of Marlborough, cost 50 guineas. It is the number and the fineness of the studs, each of which is screwed in, that enhances the value and the expence of the Woodstock polished steel articles. A pair of scissors will sell, according to the workmanship, from 5*s*. to three guineas: other articles in proportion; yet it is by no means a profitable business to the manufacturer.

About fifty years ago, the manufacture of leather into breeches and gloves began to be established here; and has gradually risen to a degree of reputation unrivalled, and to an extent which furnishes employment to the poor for many miles round. The making of breeches, indeed, from the disuse of leather in that article of dress, except among military men, has dwindled away to little; but in the same proportion the manufacture of gloves has been rising. Between 60 to 70 men are now employed as grounders of leather and cutters of gloves, who can earn from a guinea to 30*s*. weekly; and no fewer than 1400 or 1500 women and girls are engaged in making of gloves, whose wages, according to their diligence, will run from 8*s*. to 12*s*. per week.

The leather grounders have a peculiar art, in dressing the skin in such a manner as to give it at once fineness

of

of grain and tenacity of substance. The sewers too produce very durable work, and the neatness of the fabric is universally admired. Woodstock gloves likewise possess this advantage : they may be washed several times, and look well and feel comfortable to the hands till they are quite worn out. A pair of doe-skin gloves, which will cost about 5s. will last a gentleman who rides daily, nearly twelve months ; hence they are not only more elegant, but cheaper than the gloves generally made in other places.

The principal manufacturers are Mr. Cross, Mr. Dewsnap, sen. and jun., Mrs. Money, and Mr. Eldridge. The trade is extremely flourishing, and no fewer than from 360 to 400 dozen of gloves are now manufactured weekly, though ten years ago they did not exceed 20 or 30 dozen in the same space of time.

" The employment of the female poor, on the southward side of the county, is lace-making; but in the middle and northward side, the more general employment is spinning*."

———

SECT. IV.——THE POOR.

In the parish of Kempsford thirteen acres were assigned for the use of the poor, under the management of the parish officers. It was assigned in portions according to families, to do what they would with it—not lett to them, to avoid the difficulty of turning out in case of necessity : they have uniformly applied it to potatoes. The parish hired the land, and paid for it, and

---

* Original Report.

took

took the rent (the same) of the poor. The poor have found it of great benefit, and would not relinquish it on any account. But it is not made a means of lowering rates but in a very trifling degree.

Oxford Gaol.—Blankets and shags were made in it. I inquired for annual accounts, which were distributed some years ago; but the answer was—*there are no prisoners, and therefore the accounts dropped.* Good as the management was, the change is happy, and perhaps due to the war.—*See Howard on Prisons,* p. 170.

" It has been recommended* to lay to the cottages of labourers a small portion of land, in such places as might be converted into pasture; and though I must for the present produce the two following instances of the effect of such a proceeding from two neighbouring counties, yet the instances appear to me to be too much in point, to be here omitted. It is almost needless to say, the authenticity of them is unquestionable.

" The commonable land belonging to a parish in Worcestershire, which is situated very near to Tewkesbury in Gloucestershire, was enclosed about 22 years ago; and there was an allotment containing 25 acres, set out for the use of such of the poor as rented less than 10*l.* a year, to be stocked in common. At that time there were about sixteen people on the parish books, some of whom had families. Previous to the enclosure, there were some few cottages that had land lett with them, to the amount of 6*l.* or 7*l.* a year each.

---

* What follows deserves the most serious attention, and the nicest and fullest investigation; at least upon making new enclosures.—*Rev. J. Howlett.*

The

The occupiers of these cottages, with land annexed to them, were remarkable for bringing up their families in a more neat and decent manner than those whose cottages were without land; and it was this circumstance which induced the lord of the manor (to whom almost the whole of the parish belonged) to lay a plot of land, from five to twelve acres (besides the common before-mentioned), to other of the cottages, and to add a small building, sufficient to contain a horse or a cow, and likewise to allow grafting stocks to raise orchards. In some instances, small sums of money were lent to these cottagers, for the purchase of a cow, a mare, or a pig.

" The following good effects have been the consequence of this proceeding. It has not in one instance, failed of giving an industrious turn, even to some who were before idle and profligate. Their attention in nursing up the young trees, has been so much beyond what a farmer, intent upon greater objects, can or will bestow, that the value of the orchards is increased to 40*s*. per acre, in land which was of less than half the value in its former state. And the poor's-rates have, from this cause, fallen to 4*d*. in the pound, or less, there being only two (and those very old) people on the books at this time, whilst the adjoining parishes are assessed from 2*s*. 6*d*. to 5*s*. in the pound. These are labourers, and good ones; their little concerns are managed by their wives and children, with their own assistance after their day's work. Their stock consists of a cow, a yearling heifer, or a mare to breed (from which a colt at half a year old will fetch from 3*l*. to 5*l*.), a sow, and 30 or 40 geese. This, therefore, has been the means of bringing a supply of poultry and fruit to the market, of increasing population, and

making

making the land produce double the rent a farmer can afford to give.

" The other parish to which I alluded, is situated within six miles of Leicester, and belongs to a nobleman whose family have for many years lett small quantities of land, varying from four to twenty acres, with the cottages, after the rate of about one-fifth less than the same sort is lett for to the farmers. These cottagers keep from one to three or four milch cows to make butter, from five to twenty lambs (being chiefly twins purchased and brought up by hand), one or more pigs, and raise from one to three or four young beasts yearly.

" The consequence is, that about 20 families live comfortable as labourers, whilst the management of their stock employs their families, and themselves at their leisure time, which might otherwise be spent in an alehouse. The poor's-rates are only from 6d. to 8d. in the pound, which may be considered as a saving to the parish of 60l. or 80l. a year. It is true, that the landlord sustains a loss in the first instance, of about 30l. a year in rent, on account of these lands being lett cheaper than the farms; but it is doubly restored, by enabling the farmers to pay a greater rent for their farms, on account of the poor's-rates being so easy*."

" The justice, good sense, and philanthropy of the preceding, surely does infinite honour to the author, as well as to those who have set so excellent an example. Day labourers in this, and many counties, are worse off than any part of the community besides; their wages being much less in proportion ; so that, barely gaining a subsistence, and utterly unable to lay by, whilst in

---

* Original Report.

health

health and strength, they *can* have in sickness, or old age, no other resource than the poor's-rates; which frequently, by subjecting them to tyranny, and oppressive insult, stifles the conscious pride of honesty, and corrupts every good principle. And it may too frequently be observed, that the shameless effrontery of an idle and dissolute subject will *extort* relief from parish officers, whilst the patient suffering of the helpless, but *real* object of charity remains unnoticed."—(*Annotation*).

Land to Cottagers.—" I am of opinion, that if letting small portions of land with cottages was more generally in use, it would operate as a great incentive to industry, and would be a means of attaching many thousands to the present Government, who would now eagerly embrace any opportunity that would tend to the overturning it; as nothing, in my opinion, contributes more to the support of a government, than placing as large a body of the people as possible in a situation that they may have something to loose if a change should take place. Now, there being a greater number of that description of people in England, than in any other nation in Europe, the government is of course the most stable, and has less to fear from those fascinating notions of liberty and equality which now so much prevail in a neighbouring country."—*Richard Wills.*

The Poor.—" The situation and circumstances of the lower class of the community, is a subject that merits an investigation as much as any that has come under our consideration. It must be obvious to every one the least acquainted with this class of people, that
they

they are gradually degenerating in those qualities which are necessary for the attainment of their own happiness and the constituting of useful members in society. Now if this be admitted, and I am fearful it cannot be denied, I will ask what must be the consequence if it is not immediately attended to, and proper measures taken to remedy a disease of so serious a nature ? I am of opinion, that this depravity originated in the inadequacy of wages to the support of themselves and families, which compel many, contrary to their natural inclination, to have recourse to dishonest means for the attainment of that which could not otherwise be had ; and as little offences generally lead men on to the commission of greater, the mind at length becomes hardened, and instead of instilling principles of religion and morality in their children, they familiarize their tender minds to the same immoral and dishonest practices they themselves have lived in the habit of.

" It is much to be wished, as our author observes, that that laudable spirit which existed thirty years ago, could be again revived, when the poor would exert themselves to the utmost before they would apply to the parish for relief: whereas they are now not ashamed to apply upon every trifling occasion, and sometimes under feigned pretences. This very culpable conduct in the poor people, I am sorry to say, originated chiefly in past ill treatment, as it is a fact which cannot be denied, that for many years past, poor men who had large families, have performed their labour with no better a subsistence than scarcely sufficient bread and water; the whole family of course must submit to the same humiliating fare. Such distress, I conceive, will be thought sufficient to pervert a natural good disposition. And to this and other trying occasions, may

be

be attributed that propensity to pilfering, to contract-
ing of debts with an intention of never paying, to taking of every unfair advantage of their employers, and
in short, to almost every other kind of injustice in a
greater or less degree ; all of which they think their
poverty a sufficient excuse for.

" It will now be proper to consider by what means
a reformation in this class of people can be effected ;
and this I think should begin with a considerable advance of wages. If the proprietors of land had obliged their tenants to have paid their labourers 14*d.*
16*d.* or 18*d.* per day, in proportion to the size of their
families, instead of racking the rents up to a degree
which can scarcely be borne, it would have reflected
upon them immortal honour. But as this was not the
case, can it be expected that the tenants, already over-
burdened with a variety of increased expences, exclusive of a vast additional rent, will do any thing toward
an effectual relief of the distresses of the poor ?

" If an advance of wages was to take place equal to
what I have above stated, it would very materially
alleviate the distresses of many families who are now
under the necessity of procuring a considerable part of
their household necessaries by theft and plunder: the
laws for petty offences might then be more strictly
enforced, and the farmer released from those depredations which are so frequently committed upon his
hedges, turnips, corn at harvest, and many other
things in the farming business liable to be plundered ;
instead of which, he is now under the necessity of
turning his back when he sees himself being robbed,
rather than apprehend the offender and have him
brought to justice, where he knows that little or no
redress is to be had : whereas, if the poor people were
                                        placed

placed in a situation that a pecuniary fine (as a punishment) might be levied upon them, it would deter them
from committing such crimes. But now, if you take
a person to justice, who has committed any of the
above offences, the magistrate, from his known poverty,
will order a small fine to be paid; the pauper will tell
him he has nothing to pay it : here the matter ends, and
instead of a punishment, he is encouraged to persist in the
same nefarious practices, and to a still greater degree.

" Work-houses upon a liberal plan, I conceive to
be of much advantage to parishes, and conducive to
the real happiness of the poor; but it must not be of
such a nature as is usually established in country villages, where the object is merely the diminution of
expenses, without regarding the comforts of those for
whose benefit such institutions were originally intended.
Work-houses were meant as an asylum for the poor
and aged, and a place of confinement for the idle and
profligate; but being badly conducted, many who are
entitled to the benefit of such an institution, will rather
live upon a starving pittance allowed them out, than
put themselves under a savage-minded master within;
and many that it would be proper to confine, are
allowed a weekly payment out, in order to save the
trouble they would occasion within.

" I scarcely know any thing which contributes more
to the advancing the poor's-rates than a number of idle
dissolute girls, who form connexions with men as bad
as themselves, the consequence of which, is frequent
instances of bastardy. This increasing evil should be
particularly attended to, and I think might in a great
measure be remedied, by placing such girls as have
lost their father, or any others who might appear to
be destitute of a proper instructor, in a work-house,

OXFORD.]   where

where they should be brought up in a way that would render them fit for service, and when they shall attain the age of fourteen or fifteen, should be taken out by the farmers, or others who have generally occasion for such girls in the summer: they would by that means get themselves qualified for better situations, and instead of entailing upon themselves disgrace and infamy, would become useful members in society.

" I am surprised the Author has no where mentioned what nuisances so many dogs are, kept by poor people that can scarcely keep themselves, and what injury the farmer many times sustains from them*."

" It certainly is an object to every landlord, and it ought to be to every tenant, to make the situation of the poor industrious man as comfortable as possible; for we must know, that without his labour our farms would be of little value. But though every cottage ought to have land enough adjoining to it to raise vegetables, and in general has, I doubt whether it would be practicable to allow sufficient to keep a cow. Property is much divided in this part of the county, and many cottages belong to persons who have not more land than what the house stands upon, with the garden ; and I fear that it would not answer to any cottager to rent an acre or two of land of a farmer. I likewise fear, that if the quantity of land here recommended were added to each cottage, the labourer would have great difficulty to find the money to buy a cow : and if he should have the good fortune to do it for once, we know that if any accident should happen to this cow, that he must become a petitioner to raise money for a second. I mean not, however, to urge

---

* R. Wills's Note on the Original Report.

any

any thing against the interest of the poor man, but wish merely to submit to your consideration the propriety, or indeed the policy, particularly at the present time, of publishing an idea that may tend to make the industrious poor man discontented, and in consequence unhappy in his situation, which, in many instances cannot be relieved, and I believe we all find that the greatest happiness we enjoy is contentment."—*Marquis of Blandford.*

SECT. V.—POPULATION.

Population of the County of Oxford.

	Houses.			Persons.		Occupations.			
	Inhabited.	By how many Families occupied.	Uninhabited.	Males.	Females.	Persons chiefly employed in Agriculture.	Persons chiefly employed in Trade, Manufactures, or Handicraft.	All other Persons not comprized in the two preceding Classes.	Total of Persons.
Hundred of Bampton ··	2260	2581	122	5817	6253	2888	2117	7108	12,070
———— Banbury ··	1575	1700	48	3872	4108	1358	1933	4187	7980
———— Binfield ··	1392	1476	39	3135	3533	1406	893	4479	6668
———— Bloxham ··	1404	1517	26	3463	3574	1873	1026	2880	7037
———— Bullington	1340	1711	31	3688	3711	3461	837	3011	7399
———— Chadlington	2128	2630	47	5886	6049	3966	724	6163	11,935
———— Dorchester	539	605	18	1310	1339	1251	180	1205	2649
———— Ewelme ··	934	1071	33	2272	2442	2291	698	1712	4714
———— Langtree ··	632	686	10	1548	1620	1737	328	1103	3168
———— Lewknor ··	727	827	20	1795	1966	950	528	2281	3761
———— Pirton ····	510	557	5	1199	1360	818	354	1377	2559
———— Ploughley ··	2016	2270	30	4883	5010	5513	2848	1946	9893
———— Thame ····	717	794	19	1805	1896	1132	337	1973	3701
———— Wooton ··	2385	2771	53	6549	6466	3812	1175	7575	13,015
City of Oxford ·······	1827	2230	82	5920	5774	146	1647	9901	11,694
Liberty of the same ····	213	324	11	644	733	7	721	649	1377
	20,599	23,750	594	53,786	55,834	33,109	16,346	57,550	109.620

CHAP. XVI.

OBSTACLES TO IMPROVEMENT.

GRUB and Red-worm.—Around Baldon **they** are but little subject to grubs or red worms in their **corn** crops, which Sir C. Willoughby attributes to the shallow ploughing; the best farmers having a **notion** that deep ploughing encourages them, and from **the** constant folding of corn lands which keeps the **land** tight.

In grubbing the furze land of Hampton Poyle, **upon** the enclosure, Mr. Blake avoided sowing wheat, **as he** knew how apt the grub is to be in such land; **and** sowed oats.

Charlock.—" The arable lands near to the **bottoms** of the Chiltern hills, constantly produce a **considerable** quantity of charlock, but in some years more **than** others; which depends, probably, on the depth **of soil** turned up, or the rain that falls at seed time. **No other** means are used to destroy this weed, than to mow **off** the tops with a scythe, at such times as it shoots **higher** than the corn, which is not always the case; **and even** this process is often neglected, under an idea that **the** ground is so full of the seeds, that if the whole **was** pulled up one year, there would notwithstanding **be as** much the next. Having two fields of barley, and **one** of oats, in which the stems of charlock were nearly **as** numerous as the stems of the corn, I was desirous **to**

try

try the effect of weeding; and for this purpose, I pro
cured a number of hands (women and children) just
before the time of harvest, and set them to work, leaving a part of the oat-field untouched, in order to see
what would be the difference. The weather being hot,
and the charlock being spread open, and turned, in
the manner of making hay, it soon became dry, and
had a sweet smell; and when given to horses and cows,
I have several times observed both refuse very good
meadow hay, and prefer feeding on this charlock. I
I cannot ascertain the exact saving of hay by this
means; but I believe it was not less than half the expense of pulling up the charlock. Whether less will
come up on the part pulled than on that omitted, remains to be examined*."

* Original Report,

APPENDIX.

THE situation of Oxford, in nearly the centre of the kingdom, and the station of a great and flourishing University, points it out as a place peculiarly proper for any amelioration in the plan of academical studies and pursuits.

Theology and classical literature have long flourished at Oxford; and in modern times mathematics have not been neglected. Let these and all the sciences taught at our Universities expand their influence in every direction; but that Agriculture should be utterly neglected, and that Oxford and Cambridge should be the only two Universities in the enlightened part of Europe without Professors for teaching this most useful of all the Arts, is a circumstance that must excite some degree of surprise. The deficiency was felt by the late Professor of Botany in this seminary, who left a fund to be hereafter eventually applied in establishing such a Professorship.

But mere lectures are insufficient to command the attention, and give a turn to the pursuits of young men. The University abounds with those who are destined to be considerable landed proprietors, and if the proper means were used, it would not be difficult to engage them in enquiries which would form not only a most beneficial pursuit, but a rational, harmless, and entertaining amusement. The effect might be durable, and must be advantageous to the best interests of the empire.

The Ratcliff Library is endowed with 100*l.* a year salary to the Librarian, and 150*l.* a year for buying

books;

books; and as none of these have been bought for many years, there is an accumulated fund of about 5000*l*. in the trustees' hands. Whoever has received information relative to that Library, knows that the collection is a very imperfect one; and every person sees that the cases are not half filled. The vestibule of the building contains a large space and area, which is not applied to any use. The trustees are further in the possession of a large estate, applicable, as it should seem, to any public use connected with, or by construction connected with, the objects of this Library, as may fairly be conjectured from the applications which have been made of the produce: they have built and endowed the Observatory at an expense, it is said, of 30,000*l*.; and they have built an Infirmary.

To these points we may further add, the endowment with 200*l*. a year, of a Professorship of Agriculture and Rural Economy, to take place as soon as a certain botanical publication is finished; in which some progress has been made.

If these circumstances be combined, they offer an outline which might be filled up with one of the most useful and important establishments that was ever within the power of Oxford to profit by.

To buy books for making the Ratcliff a *general* Library, would at Oxford be perfectly nugatory: she possesses in the Bodleian, one of the very first class in the world, and various others; but she possesses none appropriated to Natural History, to Botany, Chemistry, Mineralogy, Agriculture, or Rural Mechanics. What I would propose (and I am supported by the respectable opinion of a most able Professor there), is to confine this Library entirely to these subjects: the funds in hand, and annual, would purchase a capital collection of books in all these branches. This would

exonerate

exonerate the Bodleian from any similar purchases, and leave their funds to Classics and Divinity. The vestibule should be a repository of Agricultural Implements; and a magnificent one it would make. The Professor of Agriculture, whenever appointed, would make a sorry figure without some such establishment; and the benefits flowing from the young men having such a means of knowledge in agricultural mechanics, with a Library rich in the most useful branches of Natural History, would have such an opportunity of acquiring those branches of knowledge most nearly connected with the cultivation and improvement of their future estates, as no place in Europe could rival. Classics and Divinity, with the other studies pursued at Oxford, are amply provided with all that can store and enlighten the mind: but when we inquire for the provisions made for those sciences immediately connected with the prosperity of the Agriculture of the kingdom, here is a dead blank: but with such an opportunity to fill it with the animated means of future improvement, as may never occur again.

I could discover but one objection to such a plan: Sir Roger Newdigate by his will left 2000*l*. for the purpose of removing the Pomfret statues, &c. to the Ratcliff Library: all that are worth exhibiting might be placed in the apertures of the circumference, and be no impediment to the plan proposed: this is not the fear: but it is to be apprehended that if the execution of the plan of decoration was begun, similar objects would, by a natural extension in the hands of artists, gradually fill all the space, and absorb all the money. Utility would fly before the schemes of elegance; and the plans of Agriculture fade away amidst the visions of *Virtu*.

I shall conclude with the words of my predecessor:

"It

" It is to the honour of this age and country, to have made numberless discoveries and capital improvements in the various arts of commerce, of civilization, and peace. If Agriculture has not hitherto been cultivated with equal success, we may hope, that under the protection of a Government, which, amid the cares of a vast empire, and the solicitude of war, consults the domestic and permanent welfare of the kingdom; under the superintendance of the Board instituted for the express purpose of Internal Improvement; and under the auspices of a Sovereign, who in this department, as in others, does not disdain to set an example to his subjects; under this propitious assemblage of circumstances, we may reasonably hope that Agriculture will soon be one of the first arts in point of perfection, as it is the foremost in point of utility and importance*."

RIVERS.

The Isis may be considered rather as a poetical than a strict appellation, for that part of the river Thames which runs near Oxford. In the old MSS.—Grants from the Crown—the river here spoken of under the title of Isis, is positively called the Thames, and I have in my possession a very old Grant from the Crown, of the Manor of Sutton to the famous Roger Mortimer, giving him a right of fishery in the river Thames, and describing its boundaries by the names which they have yet retained in the parish of Stanton Harcourt.—*Dr. Sibthorpe.*

---

* Original Report.

PLATES TO THIS WORK,

WITH VARIOUS EXPLANATIONS AND MEASURES BY
THE DRAUGHTSMAN.

Plate I. the Map.

Plates II. III. Wood-Eaton-House.   See p. 18.

Plate IV. Bishop of Durham's Cottages.   See p. 24.

Plates V. VI. The Skim-plough.   See p. 79.

Plate VII. Wiltshire Plough, used in Oxfordshire.   Made
by Mr. Pullen, at Lineham, near Wooton-Bassett,
Wiltshire.   See p. 79.

I saw this plough near Highworth, on a very high hill,
where the earth was very fleet, over rocks which frequently break the plough-shares.   There were different
kinds of ploughs at work in the same field; but the
ploughman and the farmer both spoke of this plough as
the best adapted to plough over the rocks.   M c, draws
the plough.

	Ft.	In.			Ft.	In.			Ft.	In.
a b,	6	6		t p,	0	8		vi b,	0	6
h 48,	4	0		p n,	1	8		12 11,	0	11
a h,	1	4		p 13,	1	1½		G M,	0	6
h G,	1	5		10 n,	0	10		a M,	3	3
G 16,	0	8		n 5,	2	3		M b,	3	3
G K,	1	2		44 t,	2	2		M i,	3	3
K p,	1	10		i 12,	1	0		M c,	4	5
34 b,	2	10		7 20	2	10		G c,	4	9
G 14,	1	2		i 4,	1	11		Diameter.		
t 14,	2	0		4 b,	0	4½		W,	1	8

Plate VIII.   The same.

Fig. 1.

	Ft.	In.		Ft.	In.		Ft.	In.
Fig. 1.			p 30,	2	6	11 M,	2	6
12 20,	1	8	A x,	2	6	i 6,	3	6
8 12,	1	0	x p,	1	6	4 6,	0	3
14 12,	1	2	j G,	1	6	4 M,	0	4
14 7,	0	7	30 A,	1	6	i 4,	3	3
i 8,	0	8	34 18,	1	6	8 M,	3	3
Diameter.			p t,	0	8			
W,	1	8	t 6,	0	6	**Fig. 4.**		
			v z,	0	8	13 p,	1	$1\frac{1}{2}$
Fig. 2.			o z,	0	$2\frac{1}{2}$	p 4,	0	4
a b,	6	6	30 2,	0	2	13 3,	0	3
h 48,	4	0	30 E,	1	4	13 6,	0	$6\frac{1}{4}$
48 35,	2	11	35 34,	8	7	o p,	1	8
34 b,	2	10	a M,	3	3	i o,	0	3
E a,	1	0	M b,	3	3	30 p,	2	6
a h,	1	4	M G,	0	6	p ix,	0	9
h 30,	1	$1\frac{1}{2}$				p 9,	0	9
h 16,	0	9				p 44,	2	10
1 16,	1	4	**Fig. 3.**					
16 G,	0	8	1 19,	1	7	**Fig. 5.**		
G 14,	1	2	1 11,	1	11	1 5,	1	5
G K,	1	2	11 e,	0	11	30 44,	0	9
K p,	1	10	11 9,	0	9			
			i 8,	0	8			

Plate IX. Glamorganshire Plough, used in Oxfordshire.
v v, five wood wedges of different sizes, by which they
regulate the coulter and the share.

	Ft.	In.		Ft.	In.		Ft.	In.
a b,	8	3	a 16,	1	4	1 12,	1	0
a c,	0	8	15 b,	1	3	1 x,	1	7
c b,	7	7	i f,	1	5	12 13,	3	0
c i,	3	2	p f,	1	6	k 5,	2	5
1 3,	0	3	p 17,	1	5	5 13,	1	1
c i,	3	11	p 4,	1	8	13 10,	0	10
1 i,	0	9	p t,	0	$6\frac{1}{2}$	x 12,	0	7
c 63,	5	3	4 k,	0	$4\frac{3}{4}$	k 12,	0	4
63 26,	2	2	k p,	2	$0\frac{3}{4}$	5 d,	0	9
26 29,	2	5						

Plate X. The same.
Fig. 2, the share; Fig. 5, the coulter.

Fig. 1.

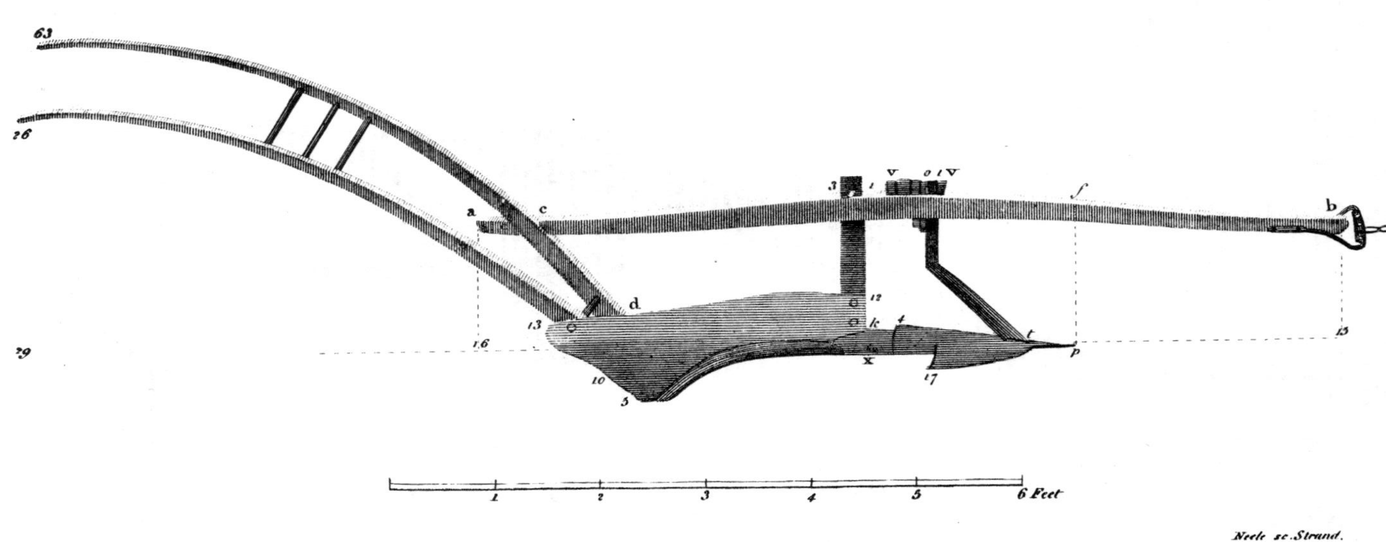

Glamorganshire Plough, used in Oxfordshire.
Ford
6 Feet
Neele sc. Strand.

Parts of the Glamorganshire Plough. 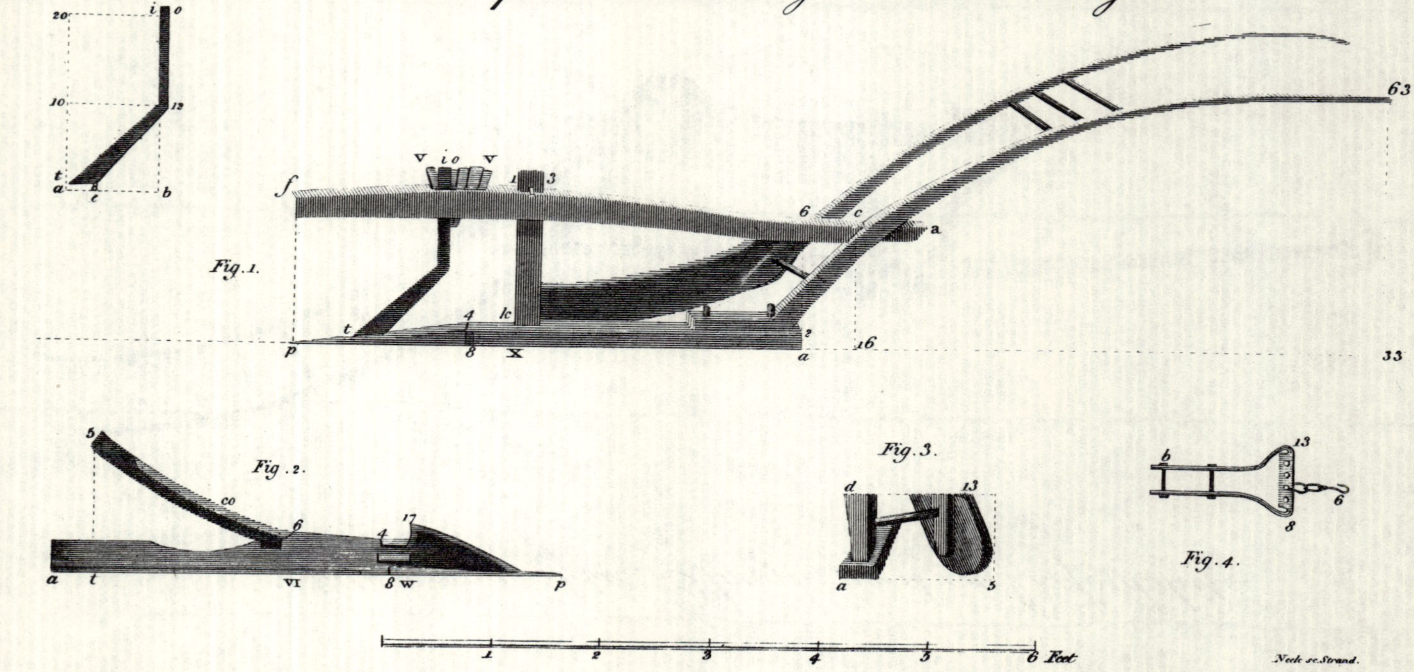
Plate X.
Fig.1.
Fig.2.
Fig.3.
Fig.4.
6 Feet
Neele sc. Strand.

	Ft.	In.		Ft.	In.		Ft.	In.
Fig. 1.			a 2,	0	2½	**Fig. 3.**		
c 63,	5	3	i o,	0	1½	a d,	0	10
63 33,	2	9				a 13,	1	3
a c,	0	8				a 5,	1	5
c 16,	1	4	**Fig. 2.**					
c 6,	0	6	a p,	4	8½	**Fig. 4.**		
a 6,	1	5	p 8,	1	8	h 13,	1	1
a 16,	0	6	p 17,	1	5	13 8,	0	8
a p,	4	8½	17 w,	0	5¾	8 6,	0	6
p f,	1	6	4 8,	0	4			
f i,	1	5	a 7,	0	4½	**Fig. 5.**		
1 i,	0	9	4 6,	0	9	i o,	0	1½
i x,	1	7	6 vi,	0	4¾	12 0,	1	0
c 1,	3	2	6 5,	2	0¾	i b,	1	8
1 3,	0	3	5 4,	2	9¾	a 20,	1	8
c f,	5	7	a 5,	1	5	20 10,	0	10
p 8,	1	8	5 t,	1	4	a 10,	0	10
p t,	0	6½	Co-con-			a b,	0	10
4 k,	0	4¾	cave,	0	1½	a e,	0	4
						i t,	2	0

Plate XI. Mr. Lumbert's Turnwrest Plough.

Fig. 2, explains its revolutions precisely: it has two shares, which, at every furrow, or every length of the field, it turns round, and turns another flag by the same as the last, laying it all the same way (after the manner they do in Kent). Fig. M N, the spindle, which is fixed in the centre of Fig. 2, as see Fig. 1.

	Ft.	In.		Ft.	In.		Ft.	In.
Fig. 1.			x i,	0	4	**Diameter.**		
u b,	6	0	36 v,	3	0	W,	3	1
c 6,	6	0	36 5,	3	6			
a E,	1	4	5 9,	0	9	**Fig. 2.**		
b 28,	2	4	v 28,	1	9	A B,	2	8
B e,	1	7	b 19,	1	7	C D,	2	8
c a,	1	6	e d,	1	10	A C,	2	1
B E,	1	4	6 h,	1	10	B D,	2	1
E M,	1	0½	6 p,	3	4	A E,	1	4
M H,	1	4	p 7,	0	7	E B,	1	4
M B,	1	0½	E T,	1	10	C F,	1	4
a 4,	2	8	a i,	4	3	F D,	1	4
4 x,	1	7	a x,	4	3	D H.	1	0½
x b,	1	9	X M,	4	1			

	Ft.	In.		Ft.	In.		Ft.	In.
H B,	1	$0\frac{1}{2}$	H M,	1	4	10 B,	2	2
A G,	1	$0\frac{1}{2}$	G 4,	0	8	22 C,	2	2
G C,	1	$0\frac{1}{2}$	4 8,	1	4	B C,	3	5
E F,	2	1	8 M,	0	8			
E M,	1	$0\frac{1}{2}$	S H,	0	8		Fig. 3.	
F M,	1	$0\frac{1}{2}$	F 22,	0	10	M N,	1	11
G M,	1	4	E 10,	0	10	i M,	0	$0\frac{3}{4}$

Plate XII.   The same, in its different position.

Fig. 3, the spring that is fixed on the handles at pins 7 p, 20 4, i h, pulls the string; K, catches the point of the share that is up, and holds it fast while the other is at work; and when that has done its furrow pull i h; let the former share B, down, and catch up the other, C; thus it will plough all one way.

	Ft.	In.		Ft.	In.		Ft.	In.
Fig. 1.			H G,	2	8	x 19,	1	5
e h,	6	0	H N,	1	4	x i,	0	4
19 a,	6	0	N G,	1	4	Diameter.		
19 b,	1	7	8 4,	1	4	W,	3	1
h 6,	1	10	G 4,	0	8			
e o,	1	10	C 13,	1	1	Fig. 2.		
h 2,	2	4	h p,	3	4	B 10,	2	2
19 28,	2	4	p 7,	0	7	B E,	1	1
19 e,	4	6	C 22,	2	2	B 13,	1	1
e C,	1	7	36 v,	3	0	10 7,	0	7
e a,	1	6	v 9,	0	6	10 14,	1	2
A C,	2	1	5 9,	0	9			
C G,	1	$0\frac{1}{2}$	e N,	1	4	Fig. 3.		
G A,	1	$0\frac{1}{2}$	N X,	4	1	7 p,	0	7
H D,	1	$0\frac{1}{2}$	a i,	4	3	7 20,	1	8
D C,	2	8	N i,	4	1			

Plate XIII. Gloucester Plough, used in Oxfordshire.

	Ft.	In.		Ft.	In.		Ft.	In.
a ix,	9	0	3 ix,	1	10	i c,	0	7
a 50,	4	2	ix 26,	2	2	e l,	0	$7\frac{1}{2}$
50 28,	2	4	31 22,	1	10	e r,	0	9
a 14	1	2	22 o,	1	10	x vi,	0	6
14 31,	2	5	i o,	0	9	i k,	3	8
31 1,	1	0	o l,	0	11	Diameter.		
31 3,	3	7	i e,	0	11	W,	1	5

Plate

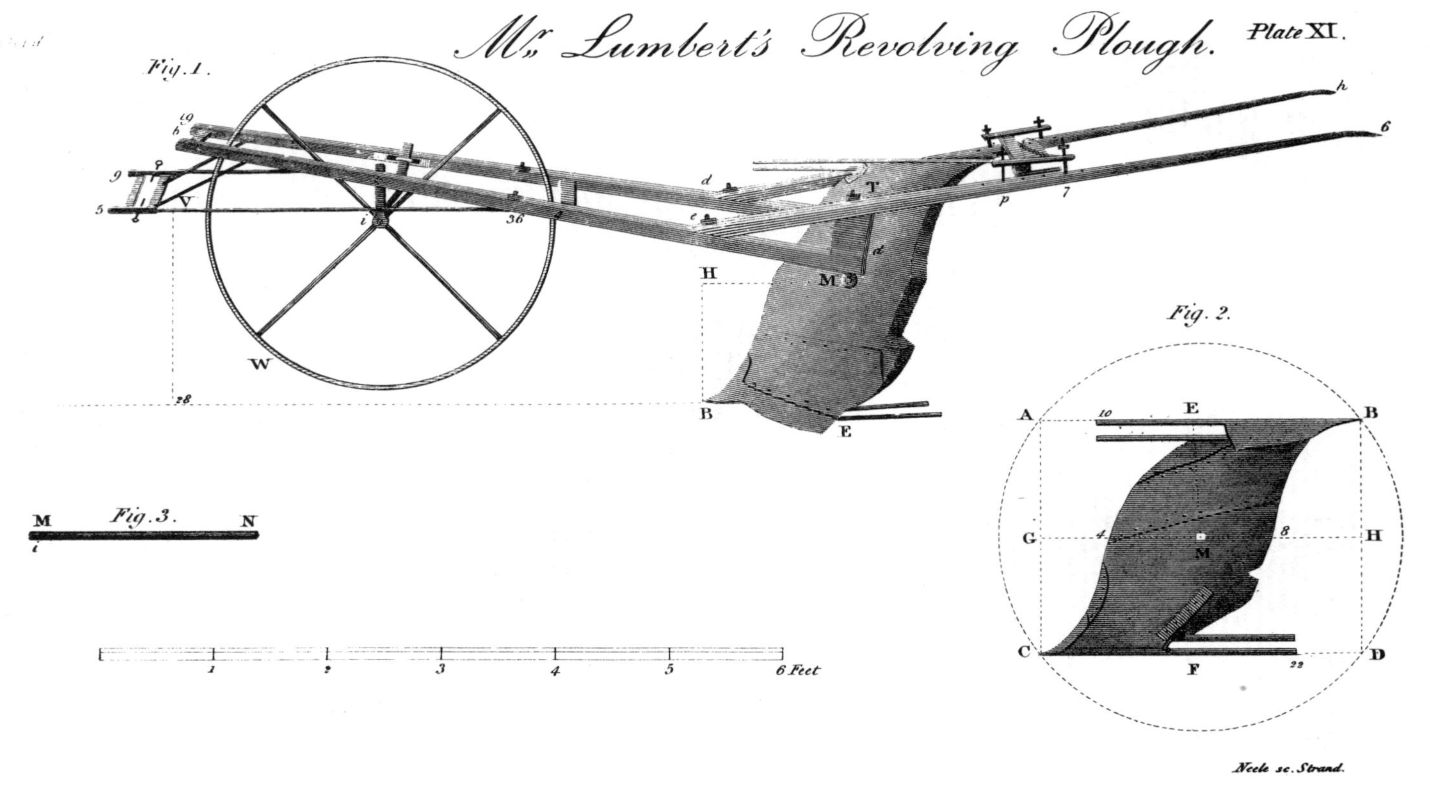

Mr Lumbert's Revolving Plough. Plate XI.
Fig. 1.
Fig. 2.
Fig. 3.
Neele sc. Strand.
6 Feet

Oxford
Mr. Lumbert's Revolving Plough.
Plate XII. 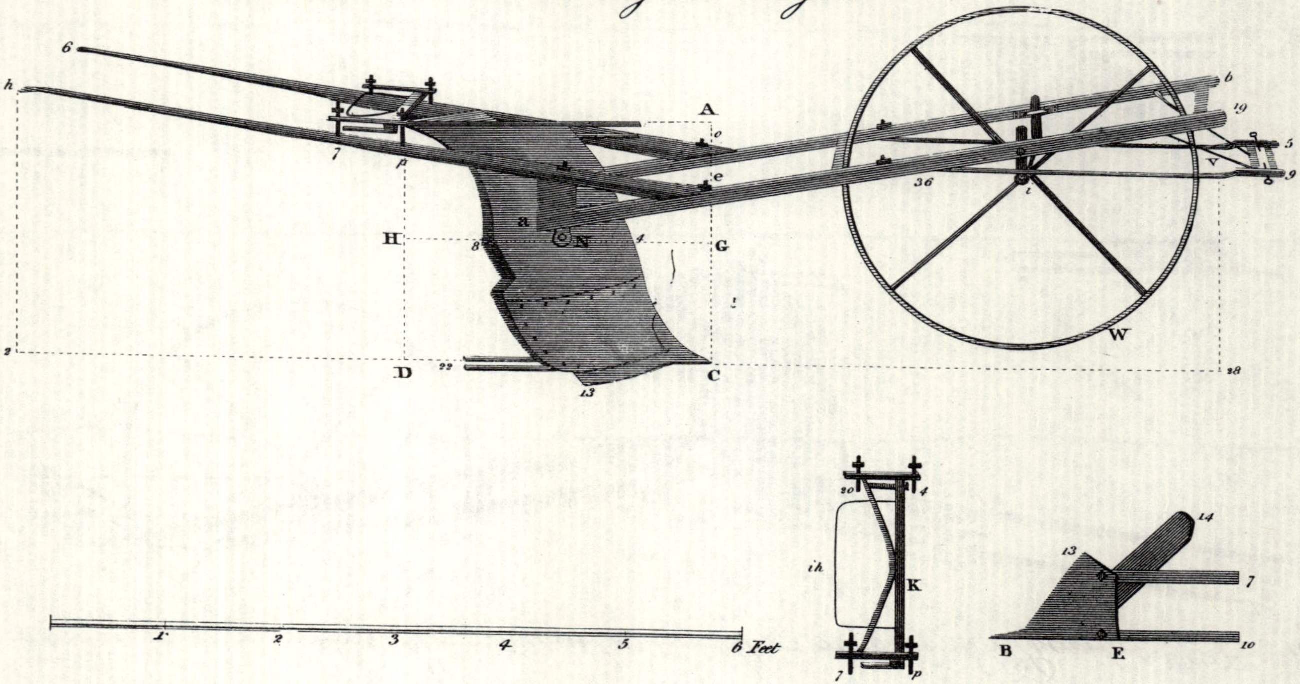
6 Feet
Neele sc. Strand.

The Common Gloucester Plough. used in Oxfordshire.

Parts of the Common Gloucester Plough.
used in Oxfordshire.
Plate XIV
Fig. 1. 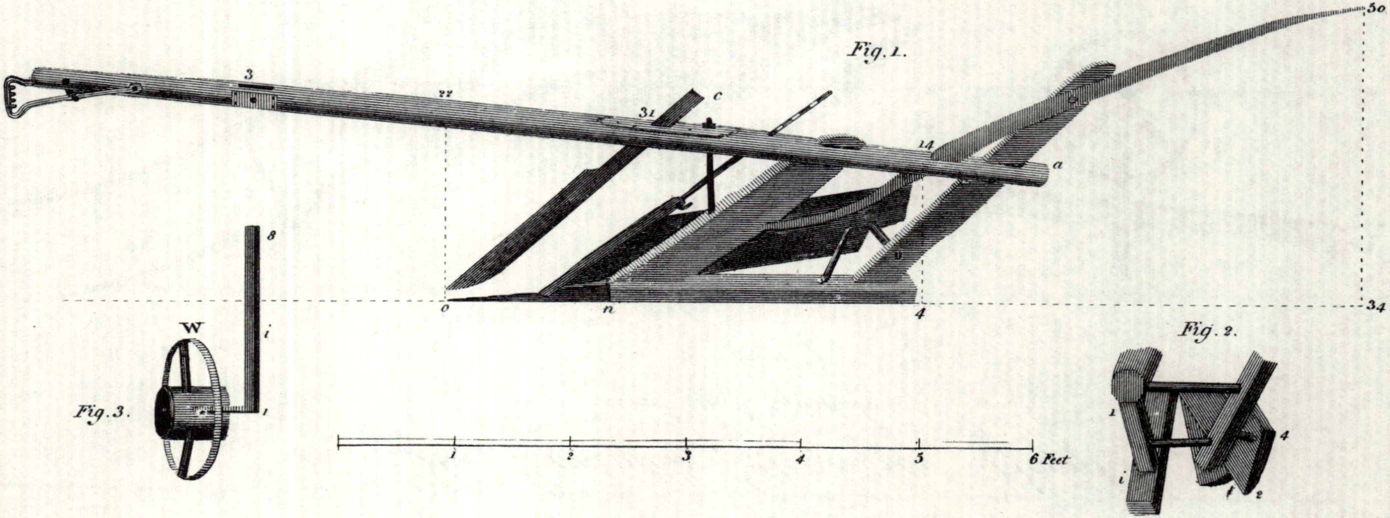
Fig. 2.
Fig. 3.
W
6 Feet
1 2 3 4 5
Neele sc. Strand.

A Double Plough used in Oxfordshire.

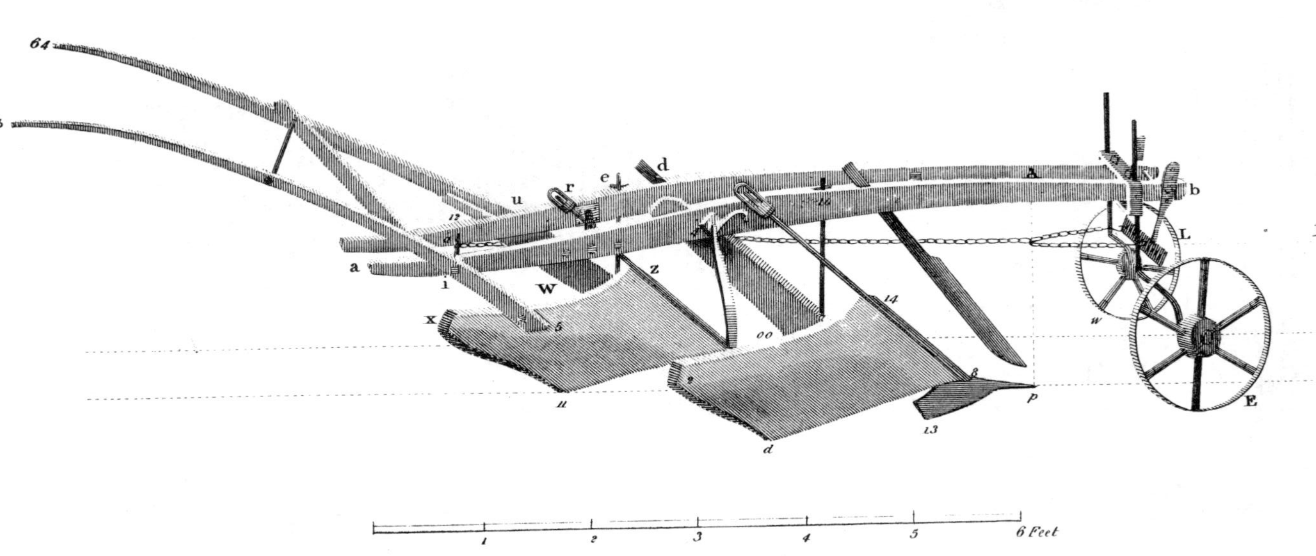

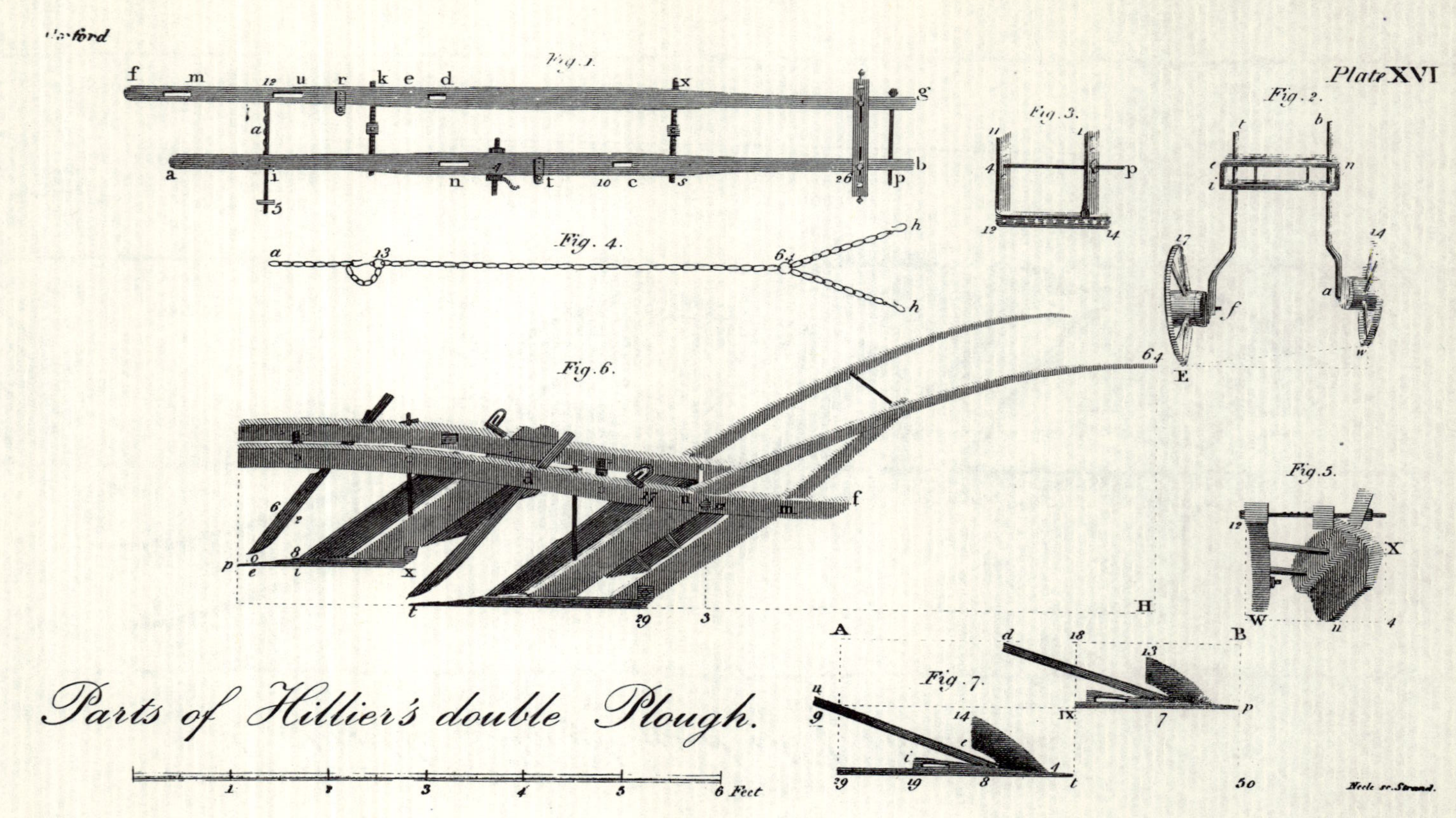

Plate XVI
Fig. 1.
Fig. 2.
Fig. 3.
Fig. 4.
Fig. 5.
Fig. 6.
Fig. 7.
Parts of Hillier's double Plough.
Neele sc. Strand.
6 Feet

Plate XIV. The same.

	Ft.	In.		Ft.	In.		Ft.	In.
Fig. 1,			n 4,	2	8	i 2,	1	2
14 50,	4	2	o 4,	4	0	i t,	0	10
50 34,	2	10	c o,	2	10			
14 4,	1	4	a 31	3	7	Fig. 3.		
a 14,	1	2	31 3	3	7	Diameter.		
14 31,	2	5				W,	1	5
31 22,	1	10	Fig. 2.			1 8,	1	8
22 0,	1	10	1 4,	1	4	i w,	0	$7\frac{1}{2}$
o n,	1	4						

Plate XV. Hillier's Double Plough, used in Oxfordshire.

The draught of this plough is very low. See the line L L, which draws by a chain from a, at the hinder part of the plough. This plough is Mr. Hillier's own invention and improvement.

	Ft.	In.		Ft.	In.		Ft.	In.
u 64,	5	4	i c,	3	9	z x,	2	0
64 b,	2	5	i 4,	2	4	14 2,	2	0
i b,	4	4	12 d,	1	10	8 d,	2	0
b 5,	5	4	d r,	1	1	W 11,	0	11
a b,	7	7	p 13,	1	$1\frac{1}{2}$	oo d,	0	11
A p,	2	0	p 8,	0	8	Diameter.		
A c,	1	5	8 14,	1	2	E,	1	5
10 c,	0	5	8 t,	2	8	w,	1	2
t c,	1	0						

Plate XVI. The same.

Parts of Mr. Hillier's Plough. Fig. 4, the chain that draws from the iron notches a, in the beam. Fig. 7, the shares: from t to 18, is one foot six inches.

	Ft.	In.		Ft.	In.		Ft.	In.
Fig. 1.			g f,	9	1	12 d,	1	10
a b,	7	7	f 12,	1	6	e d,	0	$5\frac{1}{4}$
a i,	1	1	f m,	0	8	n d,	1	1
i n,	2	1	12 u,	0	4	d x,	2	3
i 4,	2	4	12 5,	1	5	x s,	1	1
i t,	2	9	i 12,	1	0	g b,	0	$10\frac{1}{4}$
i c,	3	9	12 r,	0	9	c b,	2	9
10	0	5	12 k,	1	$1\frac{1}{4}$			

	Ft.	In.
b p,	0	3
26 c,	2	2
Fig. 2.		
f t,	2	0
a b,	2	0
b t,	1	0
e n,	1	4
e i,	0	3¼
17 14,	2	2
E w,	2	0
w 14,	1	2
E 17,	1	5
Fig. 3.		
11 i,	0	11
11 12,	1	0
12 14,	1	2
11 4,	0	4
Fig. 4.		
a 13,	1	1
a 64,	5	4
64 h.	1	4
h h,	0	10

	Ft.	In.
Fig. 5.		
4 x,	0	11
x 12,	1	5
4 11,	0	5
11 W,	0	11
Fig. 6.		
u 64,	5	4
H 64,	2	10
u f,	1	10
m f,	0	8
f 12,	1	6
12 3,	1	3
29 17,	1	5
29 3,	0	7
t 29,	2	5
t 3,	3	0
Q t,	1	10
Q d,	1	3½
p ix,	1	9
p 8,	0	8
i 8,	0	1½
e o,	0	0½
8 6,	0	6
6 2,	0	2½

	Ft.	In.
Fig. 7.		
29 6,	2	5
t 14,	1	2½
1 4,	0	4
4 14,	0	10¼
14 e,	0	3¾
14 8,	0	8
t 19,	1	7½
19 i,	0	3
e 11,	1	8
29 11,	0	11
29 9,	0	9
ix p,	1	9
p 13,	1	1½
1 13,	0	3½
13 7,	0	7
1 d,	1	6
p d,	2	6
t 11,	2	10
t 50,	1	9
t ix,	0	9
ix 18,	0	9
B 50,	1	6
A B,	4	2
29 50,	4	2

Plate XVII. Mr. Estcourt's Bean Dibbler.

With this machine, six quarts set on one acre produced 40 bushels of wheat ; and 12 quarts of barley on one acre produced 52 bushels.

	Ft.	In.
Fig. 1.		
dia. W,	4	0
e 10,	10	0
e x,	2	11
e 3,	3	0
3 4,	3	10
i n,	4	9
d 30,	2	6
xi 66,	5	6
11 22,	1	10
xi 11,	1	10
22 66,	1	10

	Ft.	In.
66 44,	0	11
44 22,	0	11
xi d,	0	11
d 11,	0	11
d 44,	3	8
Fig. 2.		
x 9,	1	9
d 30,	2	6
i d,	0	7½

	Ft.	In.
Circumf.		
	7	6
Fig. 3.		
4 9,	0	6
o 4,	0	4
i o,	0	4
Circumf.		
i 4,	0	11
Fig. 4.		
c S,	2	6

Plate

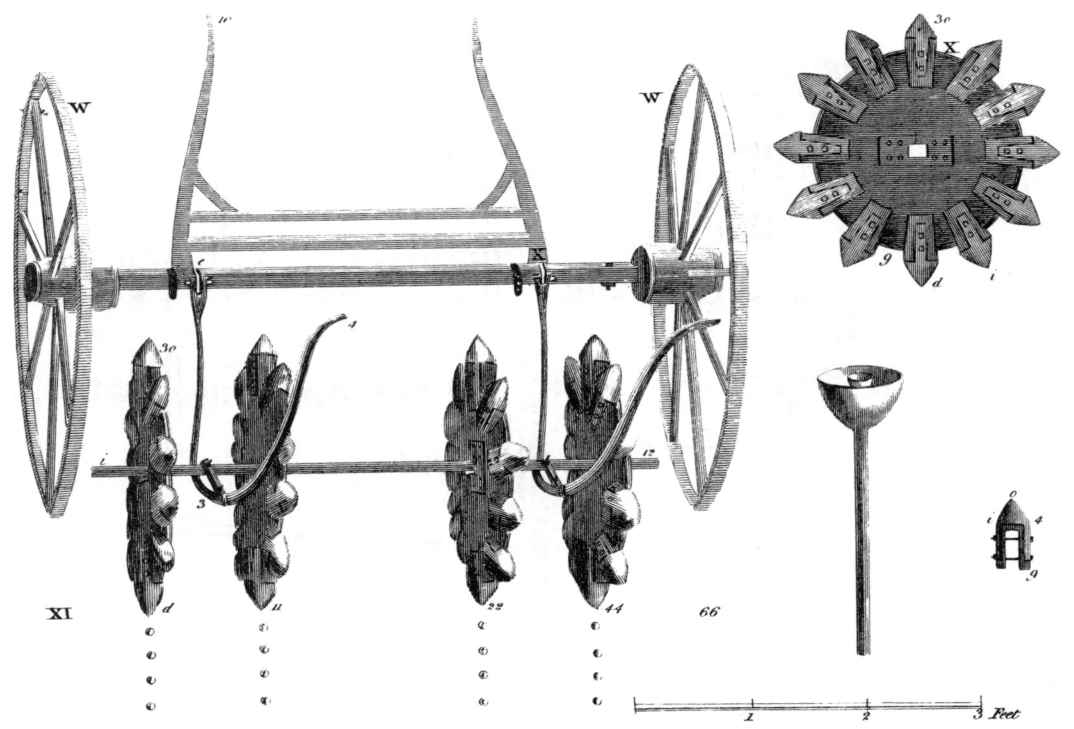
W
W
X
XI
Feet
1   2   3 Feet

PL. XVIII
Mr. Emerson's Cultivator.
Neele sc. Strand.
3 Feet

Plate XVIII. Mr. Emerson's Cultivator.

	Ft.	In.		Ft.	In.		Ft.	In°
a b,	7	1	n f,	1	4	N N,	2	6
x B,	5	2	74 30,	2	6	b 18,	1	6
x 4,	1	1	a 9,	0	9	i p,	1	6
v 4,	0	4	o Y,	3	2	iv b,	4	6
n b,	2	6	Y S,	1	6	p t,	4	6
n R,	2	4	W 17,	2	3	x 8,	1	9
5 h,	3	7	E a,	2	3	Diameter.		
h 25,	2	1	e 4,	1	11	t 8,	1	6
T 28,	2	4	ee 41,	2	1	W,	1	6
T 74,	6	2	a k,	3	6			

Plate XIX. The same.

Parts of Mr. Emerson's Complete Cultivator. Fig. 1, represents the beam, and its irons fixed proper for fixing the shares in ; W W, places to fix the wheels, as commonly used, with the seven shares. The first drawing shews it complete, as at work ; and Fig. 7, in this drawing explains how to set the shares exact. It may at pleasure be varied in the number of shares : sometimes to use it with three or five, as see Fig. 5. When the large share is used, if you wish to make ridges, put the large share Fig. 4, in at R, in Fig. 1, and take two of the fore shares out, and remove the other two shares to N N, and as you may clearly see in Fig. 5. how to set them, with the great share following them. But observe, this share is never used in Fig. 7, which is the way that Mr. Emerson generally uses it with six horses (put double); and when the great share is used, remove the wheels to W h W h. Fig. 6, represents a section of the shares as they should be set to work on Mr. Emerson's principle, not on a plane board, as has been the general mode of setting tools to work, as is seen by the Figure I have shewn : each share is set on a different plane line, besides having each an inclination to point downwards. See the line E ee, the earth; p, the point of the fore-share, on the line 2, which is two inches deep; 14, the point of the second or follow‚ ing share on the line 4; which is two deeper, making it four inches deep, and thus much of this Figure represents the section of setting the shares as most used. But when the great share is added, its point e, is to be set on the line 5, which makes it five inches deep.

N. B. The wheels to be put as high as the ground line E in Fig. 6, but may be altered by the notches in Fig. 2, and Fig. 3. See also the same in the first drawing, and observe, the higher the wheels are fixed the deeper the shares will plough the ground.

Twice ploughing with this machine prepares the field fit for drilling, for which it appears particularly adapted, as it makes the ground much finer than the common modes of ploughing; besides, in seed-time its expedition may be very useful, as at once it ploughs seven times as much as the common plough of nine inches.

The use of the handle T, (in the first drawing) is to turn 74 W, the fore wheels, when wanted to return the machine; thus lift the handle T, out of the iron fork f, and the fore-wheels will turn to the right or to the left as may be wanted, and when at work, the handle T, resting in the fork f, keeps the machine straight forward, and to remove it to another field, let the wheels down by notches, to keep the shares up.

	Ft.	In.		Ft.	In.		Ft.	In.
Fig. 1.			16 41,	1	2	Fig. 2.		
x B,	5	2	16 20,	1	8	74 c,	1	2
13 B,	1	1	v iv,	0	7	c 12,	1	0
4 x,	1	1	o o,	0	4	c 16,	1	4
n b,	2	6	b 18,	1	6	74 30,	2	6
b 5,	0	5	18 36,	1	6	74 E,	3	3
n 14,	1	2	36 iv,	1	6	W W,	2	1
14 R,	1	2	iv b,	4	6	W 18,	1	6
x 12,	1	0	N N,	2	6	18 24,	2	0
x 16,	1	4						

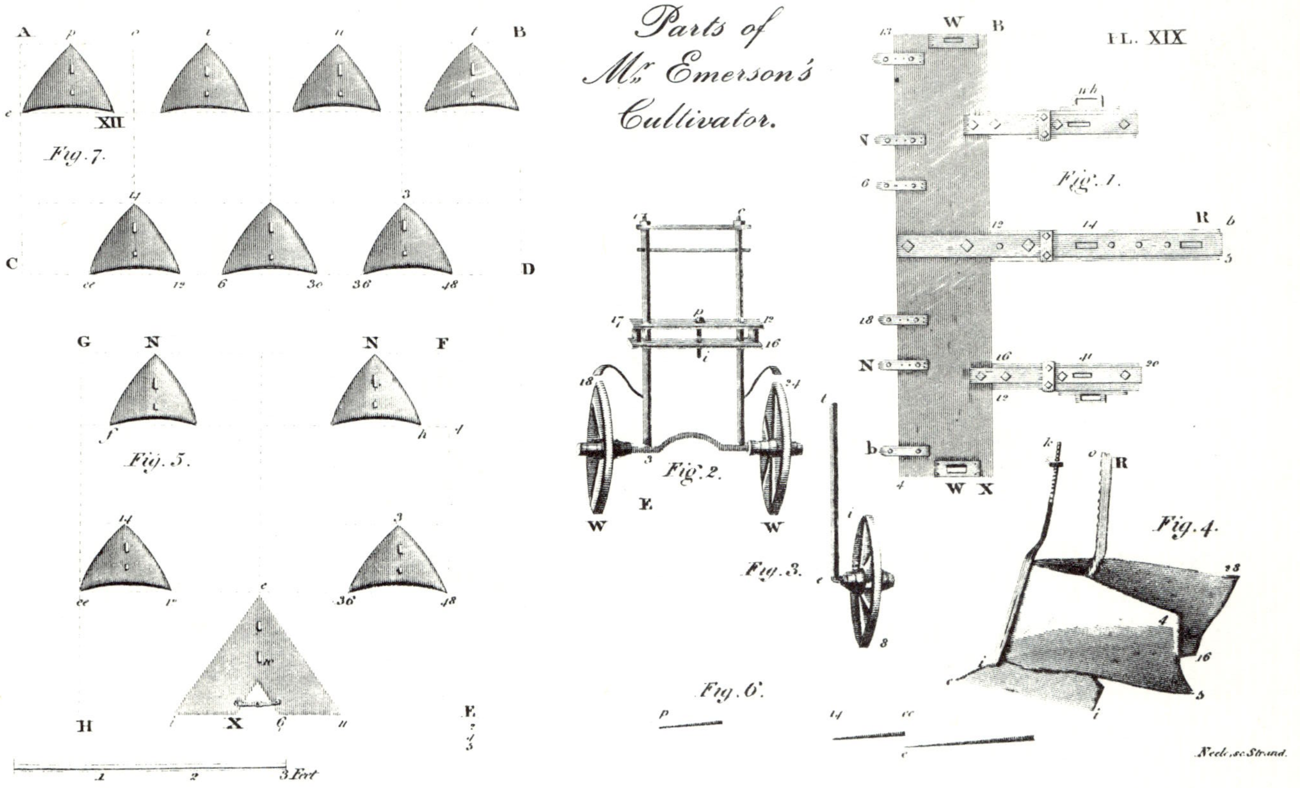

Parts of
Mr. Emerson's
Cultivator.

PL. XIX

Fig. 1.

Fig. 2.

Fig. 3.

Fig. 4.

Fig. 5.

Fig. 6.

Fig. 7.

Neele sc. Strand.

Mr. Emerson's Drill.

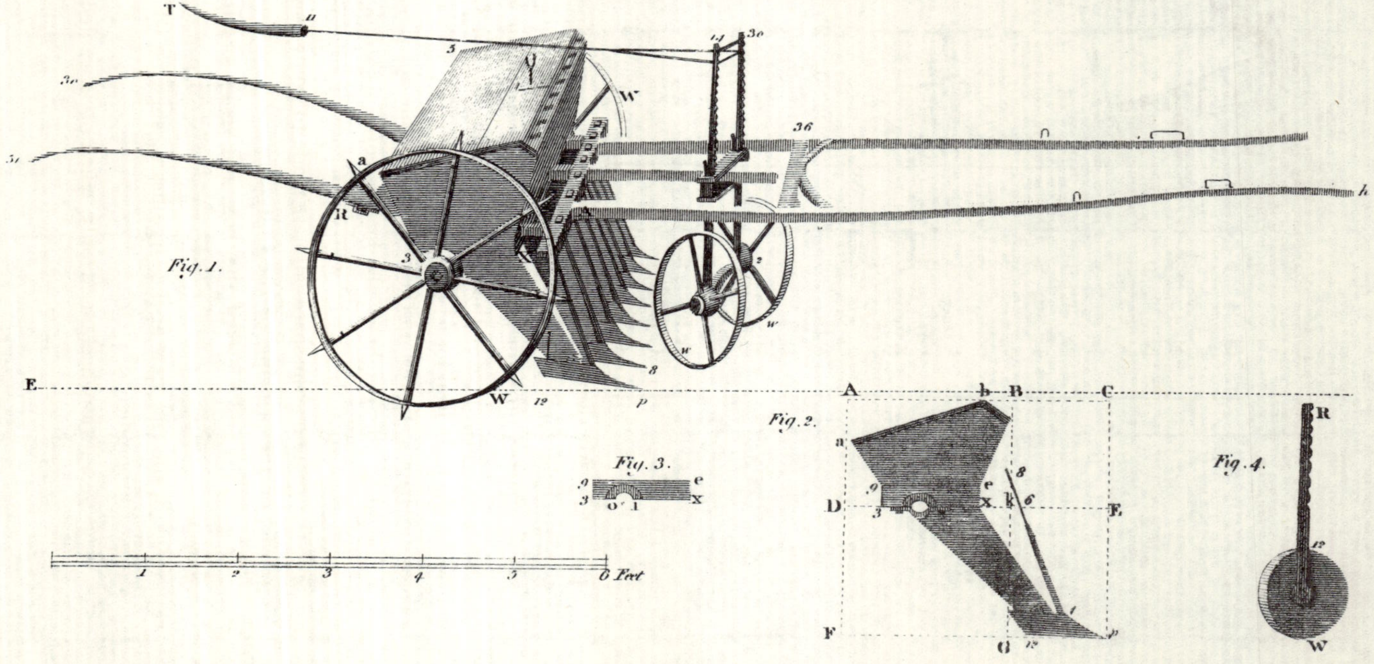

	Ft.	In.		Ft.	In.		Ft.	In.
12 17,	1	6	G H,	4	3	A e,	0	10
i p,	0	6	F 48,	2	10	A p,	0	6
			48 4,	2	0	p o,	0	9
Fig. 3.			N F,	0	9	o i,	0	9
e t,	2	0	N N,	2	6	i n,	1	6
W 8,	1	6	14 3,	3	0	n t,	1	6
i w,	0	2	ee 12,	1	0	e p,	1	0
			12 36,	2	0	p xii,	1	0
Fig. 4.			e 1,	0	4	e xii,	1	0
e i,	1	8	e 10,	0	10	Con-cave,	0	1
e 1,	0	4	e i,	1	8			
1 2,	1	2	i x,	0	$8\frac{1}{2}$	xii 14,	1	2
1 k,	2	8	x 6,	0	6	14 12,	1	0
k R,	0	$6\frac{1}{2}$	i 11,	1	11	12 ee,	1	0
o R,	0	2	ee e,	2	0	12 6,	0	6
2 4,	2	0	e 48,	2	0	ee 30,	2	6
4 28,	2	4	f h,	3	6	30 36,	0	6
28 16,	1	4	ee 48,	4	0	ee 48,	4	0
16 5,	0	5				14 3,	3	0
1 5,	2	3	**Fig. 7.**			o 14,	2	0
			A B,	5	6	c ee,	0	9
Fig. 5.			C D,	5	6	p t,	4	6
G F,	4	0	A C,	2	10	e x,	5	6

Plate XX. Mr. Emerson's Drill.

Fig. 1, the fore, or turn wheels, are the same sort as his Complete Cultivator ; their use also to return the machine, as represented in this with the handle T, lifted up out of the little iron fork f ; but to work straight forward, let the handle T, rest in the fork f, and the notches are to regulate the height of the fore-wheels ; according to which regulation the shares will be deeper or otherwise into the ground ; the shares also are made to point downwards, on the principle of the Cultivator.

Fig. 2, the end view of the box and share precisely shewn.

Fig. 3, the end view of the large beam, shewing where to fix the spindle.

Fig. 4, a little wheel, used only when it is wanted to

remove

remove the drill to another field: to keep the shares and seed-wheels up from the ground, one is to be fixed at each handle, see R, and at the same time let the fore-wheels down, and the shares and seed-wheels not touch the ground. Fig. 1, represents it as used in the field (Mr. Emerson showed me some of his fields of wheat drilled with this machine); I never saw better drilling; it was very regular, and looked beautiful, both as a specimen of his Drill, and also of his Complete Cultivator.

	Ft.	In.			Ft.	In.			Ft.	In.
Fig. 1.				**Diameter.**				k E,	1	2
51 30,	2	6		W,	2	10		E p,	1	5
E 51,	2	6		w,	1	6		G k,	1	5
51 R,	3	6						G p,	1	2
51 3,	4	3		**Fig. 2.**				p 12,	1	$0\frac{1}{2}$
A h,	8	6		A C,	2	$11\frac{1}{4}$		x p,	2	0
A H,	2	6		F p,	2	$11\frac{1}{4}$		p t,	0	7
H 36,	3	0		A F,	2	9		t 6,	1	3
2 30,	2	6		C p,	2	9		t 8,	1	8
30 14,	1	2		A B,	1	$9\frac{1}{4}$		x 6,	0	6
14 n,	4	4		A D,	1	4		a 1,	1	0
n T,	1	4		D 3,	0	$5\frac{1}{4}$				
14 T,	5	8		3 x,	1	1		**Fig. 3.**		
A 3,	1	7		x e,	0	3		e x,	0	3
x 3,	1	1		e 9,	1	1		x 3,	1	1
3 a,	1	0		9 3,	0	3		3 o,	0	5
a 5,	5	5		9 a,	0	9		i x,	0	7
x p,	2	0		a b,	1	7		o i,	0	1
p 12,	1	$0\frac{1}{2}$		b F,	0	$2\frac{3}{4}$		**Fig. 4.**		
p 8,	0	8		f e,	0	11		W R,	2	6
				x k,	0	3		W 12,	1	0

Plate XXI. The same.

Parts of Mr. Emerson's Drill. Fig. 1. a front view of the seed boxes, shewing the eight regulators.

Fig. 2, the large beam, where the seed casting wheels work under the boxes.

Fig. 3, N 7, the spindle; t W, the large seed wheels; E E, the seed casting wheels; K 4, the seed.

Fig. 4,

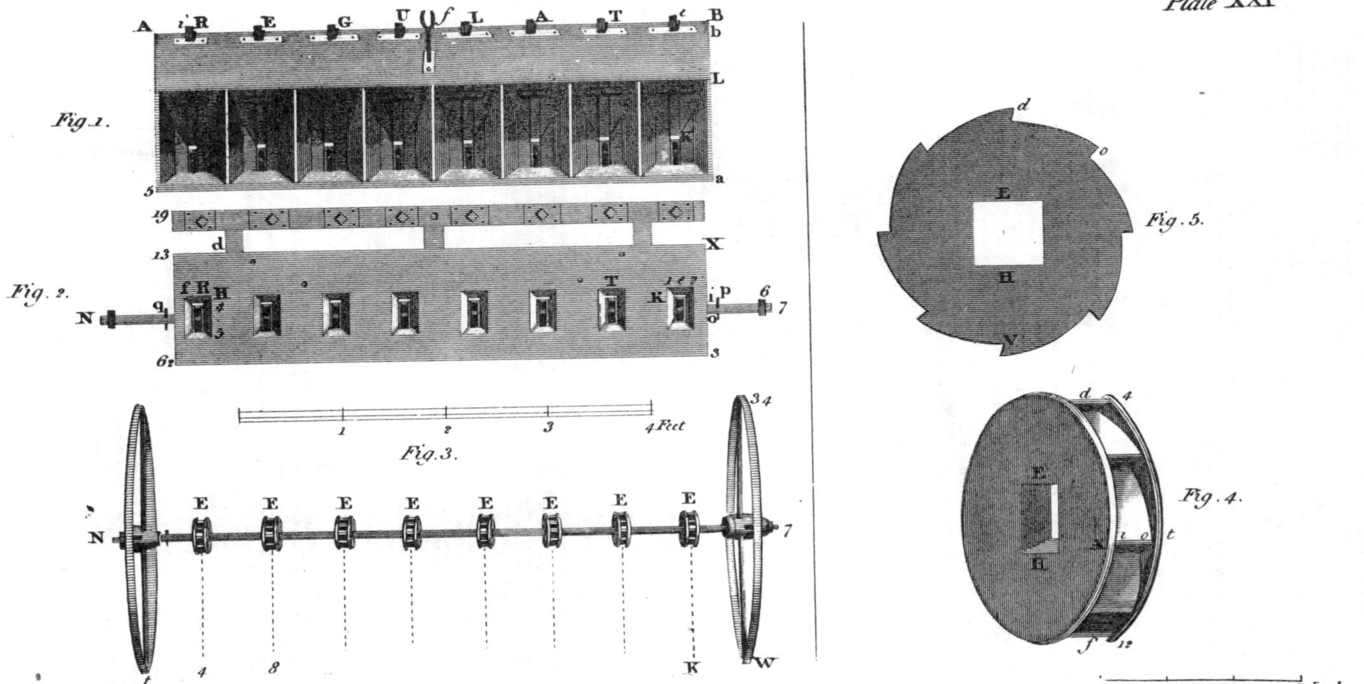

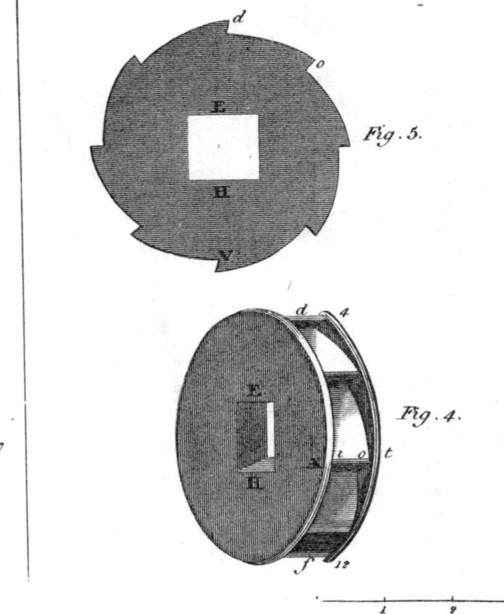

Parts of Mr. Emerson's Drill.

A Draw hoe &c. Mr. Emerson.

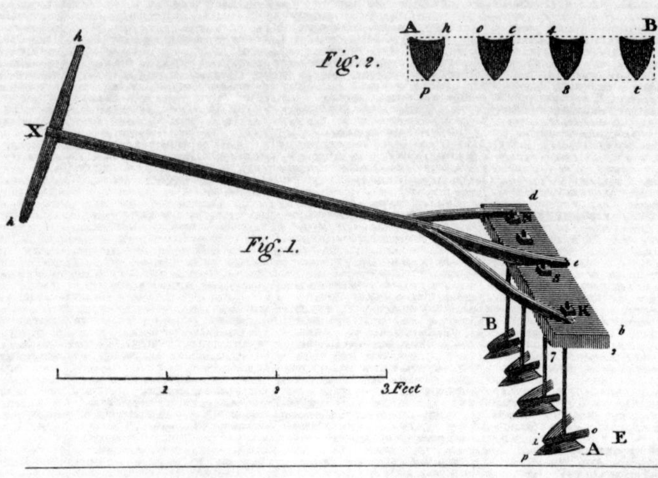

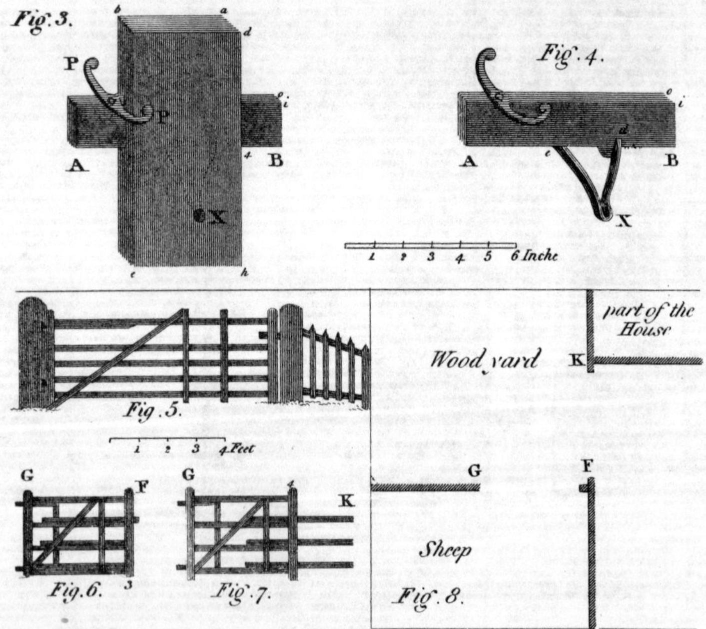

Fig. 4, the seed casting wheel, the same as E, on a larger scale.

Fig. 5, a section of the same seed casting wheel without its sides, explaining its notches and their inclination precisely, and each notch casts some seed upon every twelve inches and three quarters of ground, its revolution being eleven inches and five-eighths, and the revolution of each of the large seed wheels that mark the ground is eight feet six inches. It sows five feet four inches at once.

Fig. 1.	Ft.	In.	Fig. 2.	Ft.	In.	Fig. 3.	Ft.	In.
A B,	5	5	19 62,	1	7	N 7,	6	6
B b,	0	$2\frac{3}{4}$	62 13,	1	1	W 34,	2	10
a b,	1	7	62 3,	5	2	W t,	5	11
a L,	1	0	3 x,	1	1	t 4,	0	$7\frac{1}{2}$
a 5,	5	5	o 3,	0	5	K W,	0	$7\frac{1}{2}$
A f,	2	$8\frac{1}{2}$	i 3,	0	4	K 4,	4	8
f B,	2	$8\frac{1}{2}$	3 x,	1	1	4 8,	0	8
A i,	0	$3\frac{7}{8}$	i p,	0	1			
i e,	4	$9\frac{1}{4}$	i 6,	0	6	Fig. 4.		
i R,	0	$1\frac{1}{4}$	i 7,	0	7	Diameter.		
R 4,	1	7	P q,	5	4	E,	0	4
R E,	0	8	N q,	0	6	12 4,	0	4
E G,	0	8	N 7,	6	6	s d,	0	$3\frac{7}{8}$
G U,	0	8	f H,	0	$2\frac{3}{4}$	E H,	0	1
U L,	0	8	H d,	0	5	A t,	0	1
L A,	0	8	H 5,	0	5	i o,	0	$0\frac{5}{8}$
A T,	0	8	5 4,	0	4			
T e,	0	8	1 2,	0	$2\frac{3}{4}$	Fig. 5.		
e K,	1	7	1 K,	0	1	s d,	0	$3\frac{7}{8}$
R e,	4	8	e T,	0	8	v s,	0	$0\frac{1}{5}$
			R e,	4	8	o d,	0	$1\frac{18}{40}$

Plate XXII. Emerson's Draw-hoe, &c.

Fig. 1, Mr. Emerson's hand-hoe for drilled wheat: the points of these hoes should incline downwards, on the same principle as Mr Emerson's other machines: one man uses this machine; it has four hoes fixed proper for hoeing wheat drilled with his drill, and at once it hoes exactly

actly half the width of the drill; thus the drill and cultivator correspond ; the drill sows 64 inches, and the cultivator ploughs 63 inches at once.

Fig. 2, explains how to set the hoes.

Fig. 1.

	Ft.	In.
x e,	5	2
h h,	2	0
a b,	2	10
b E,	1	2
b 2,	0	2
2 7,	0	7
K N,	2	0
K 8,	0	8
p t,	2	0
A B,	2	4
i o,	0	$4\frac{1}{2}$
o v,	0	$3\frac{1}{2}$
A p,	0	5

Fig. 2.

	Ft.	In.
A p,	0	5
p h,	0	5
A h,	0	4

	Ft.	In.
h o,	0	4
o e,	0	4
e 4,	0	4
8 t,	0	8
p t,	2	0
A B,	2	4

Fig. 3.

	Ft.	In.
e h,	0	4
h d,	0	8
a d,	0	2
A B,	0	8
h 4,	0	4
4 B,	0	$1\frac{3}{4}$
i B,	0	$1\frac{3}{4}$
i o,	0	$0\frac{3}{4}$
4 x,	0	$2\frac{3}{4}$
P p,	0	6

Fig. 4.

	Ft.	In.
A B,	0	8
B i,	0	$1\frac{3}{4}$
i o,	0	$0\frac{3}{4}$
B e,	0	$4\frac{1}{2}$
e d,	0	$2\frac{1}{4}$
d x,	0	$3\frac{1}{8}$

Fig. 6.

	Ft.	In.
G F,	4	0
F 3,	3	0

Fig. 7.

	Ft.	In.
G 4,	4	0
G K,	6	0

Fig. 8.

	Ft.	In.
K F,	4	0
G F,	4	0

Plate XXIII. The Woodstock Waggon.

Plate XXIV. The same.

Fig. 1.

	Ft.	In.
79 69,	12	4
12 9,	12	4
79 12,	1	0
69 9,	0	9
9 15,	1	3
15 L,	5	3
L X,	5	9
X 4,	4	3
4 N,	5	9
e x,	6	0
e d,	11	4
e i,	0	10
i o,	0	6

	Ft.	In.
i g,	0	11
g j,	4	4
R 22,	1	10
R 14,	1	2
R B,	1	0
x N,	0	11
N K,	0	4
M N,	4	1
L M,	0	4
L K,	3	5
h d,	3	8
G F,	2	0

Fig. 2.

	Ft.	In.
a b,	12	4
D 72,	6	0
i D,	0	10
e o,	5	4
i i,	4	4
x 4,	4	3
15 40,	3	4
L K,	3	5
x L,	5	9
x 15,	11	0
x E,	4	0
E t,	5	4
62 79,	5	8

The Woodstock Waggon.

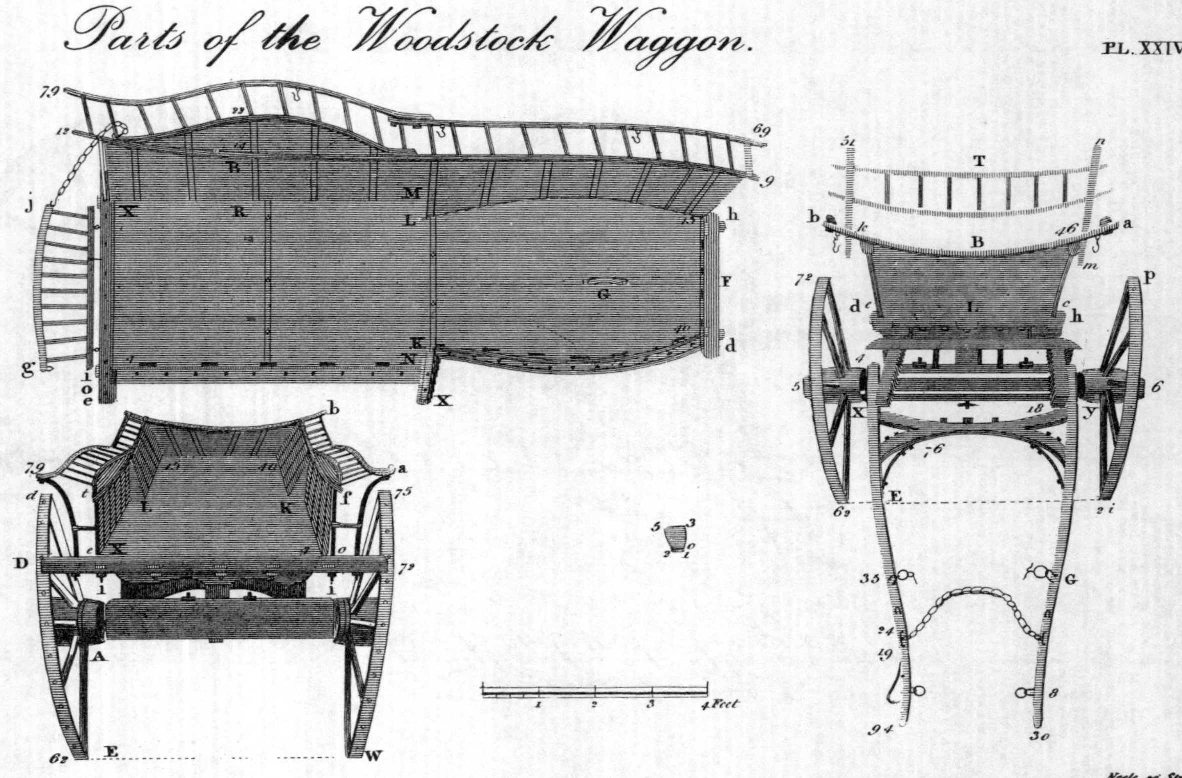

Parts of the Woodstock Waggon.
Oxford
PL. XXIV
Scale of Feet

	Ft.	In.		Ft.	In.		Ft.	In.
t f,	4	7	94 x,	7	9	a b,	5	4
a 79,	6	7	x y,	3	8	E d,	3	10
75 d,	6	5	35 G,	3	3	2 i,	0	$2\frac{3}{4}$
62 W,	5	2	94 30,	2	6	5 6,	6	0
Diameter.			94 76,	6	4	72 q,	6	0
62,	5	3	h 18,	1	6	i 62,	5	2
W,	5	3	h d,	3	8	i q,	4	4
A,	1	0	e c,	3	4			
			k 46,	3	10	Fig. 4.		
Fig. 3.			e k,	1	5	i o,	0	$0\frac{3}{4}$
4 94,	8	2	L B,	1	2	i 2,	0	$2\frac{3}{4}$
94 19,	1	7	B T,	1	8	5 3,	0	3
94 24,	2	0	m n,	2	0	2 5,	0	5
94 35,	2	11	51 n,	4	3			

Plate XXV. Ox-Stalls.

Fig. 1, a ground plan of Mr. Ruck's ox-stalls. A B, 100 feet to tie up two oxen at a stall; it holds 20 oxen. There is a door or window opposite each pair of beasts. M, the manger; P, the pump; F C D, twelve open stalls for twelve beasts loose; E F, 100 9, is a paved passage nine feet wide, for convenience of attending on the cattle; G, the gate to the ox-stalls.

Fig. 2, a specimen of the manger; 4 x, the water-troughs; W t, the channel for conveying water to all the troughs; i 20, the chain; k 20, the leather strap to buckle upon the neck of the ox; N K, the neck-rail; i. e. the distance between the manger and the neck-rail, and according to this plan, the oxen are prevented from goring each other with their horns.

Fig. 3, a small cart which will carry from 3 to 4 cwt. of hay, with which a boy may fodder the cattle, running the cart along the passage.

Fig. 5, the wall for the ox-house, three feet high stone, and then abated to a nine-inch brick wall. 3 y, is six inches of free stone.

Fig. 1.

	Ft.	In.		Ft.	In.		Ft.	In.
Fig. 1.			i 16,	3	6	7 4,	4	0
A B,	100	0	10 16,	16	0	4 5,	5	0
C D,	100	0				R e,	5	0
A D,	48	0	Fig. 2.			e 5,	2	11
B C,	48	0	N K,	10	0	4 2,	2	0
C 12,	12	0	i e,	2	7	a b,	1	11
12 F,	11	0	i t,	1	6	a E,	1	6
B E,	16	0	i 4,	3	0	Diameter.		
E F,	9	0	4 x,	10	0	W,	2	4
C F,	23	0	i 20,	1	8			
c 8,	0	8	20 k,	1	8	Fig. 4.		
B 10,	10	0	h d,	3	4	14 6,	6	0
10 x,	5	6	i h,	8	6	14 3,	3	0
x 4,	4	0				3 y,	0	6
4 i,	3	0	Fig. 3.			14 E,	1	2
i t,	1	6	a 7,	7	0	6 ix,	0	9

Plate XXVI. Mr. Emerson's Draining Plough.

It is drawn by six horses put double, as the whipple-trees shew : put the hooks h, on at e, in Fig. 1, by which the shortness of the draught is seen. The back of the coulter is half an inch thick.

N.B. The principle on which this is set to work is the same as all Mr. Emerson's machines, which is the reason they work so effectually ; see the inclination of the share c p, & & in the drawing.

	Ft.	In.		Ft.	In.		Ft.	In.
Fig 1.			a D,	1	$8\frac{1}{2}$	p 14,	1	2
a 4,	4	0	e Q,	1	$8\frac{1}{2}$	c xi,	0	11
a F,	1	7	a E,	0	$6\frac{1}{2}$	x v,	0	$3\frac{1}{4}$
F h,	4	1	i t,	2	2	v i,	0	$4\frac{1}{2}$
b h,	1	11	f 2,	0	4	i p,	0	$5\frac{1}{2}$
H h,	2	6	2 m,	0	6	o x,	0	$2\frac{1}{3}$
16 20,	1	8	m N,	0	$4\frac{1}{2}$	c p,	1	$1\frac{1}{4}$
a f,	1	8	N 14,	0	2	c q,	0	3
a 2,	2	0	m p,	1	$8\frac{1}{2}$	d D,	0	$4\frac{1}{6}$
2 4,	2	0	14 xi,	1	$1\frac{1}{4}$	E D,	1	2
4 e,	0	2	o p,	1	$1\frac{1}{4}$	Circumf.		
y z,	0	$6\frac{1}{4}$	K q,	1	4	x,	0	7
e 11,	0	11	N p,	1	4	v,	0	$6\frac{1}{2}$
e a,	4	2	c K,	1	1	i,	0	5
						Fig. 2.		

Mr. Ruck's Ox-Stalls.

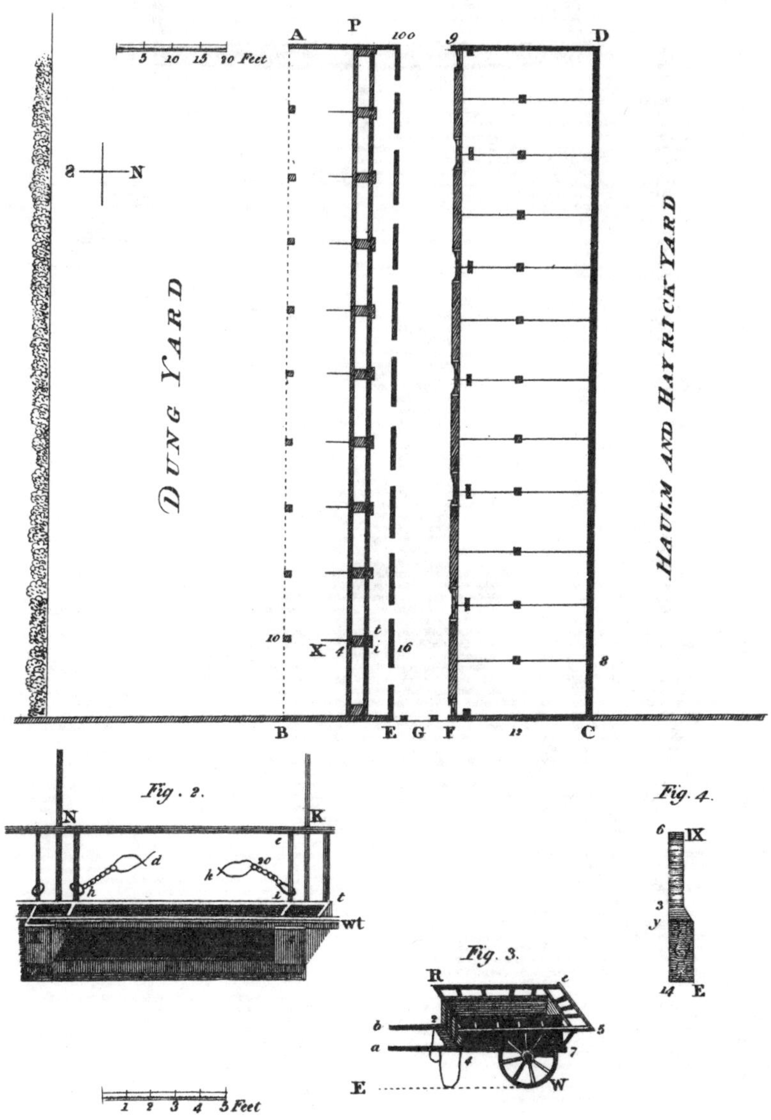

Mr. Emerson's Draining Plough.

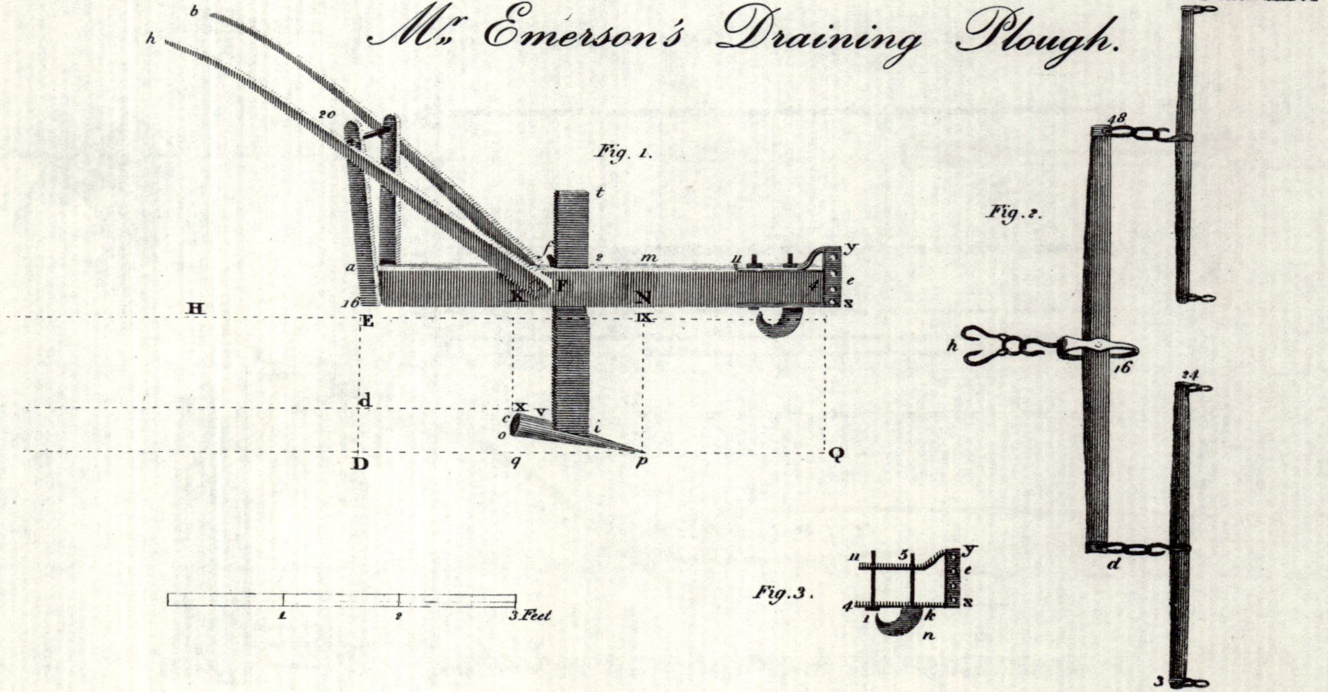

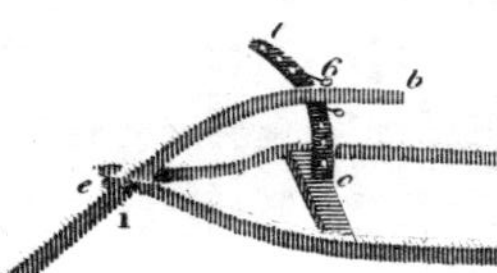
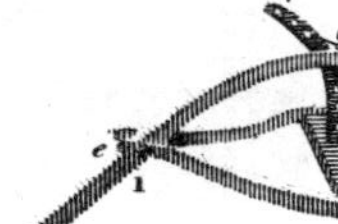
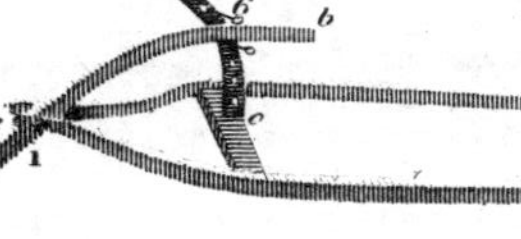
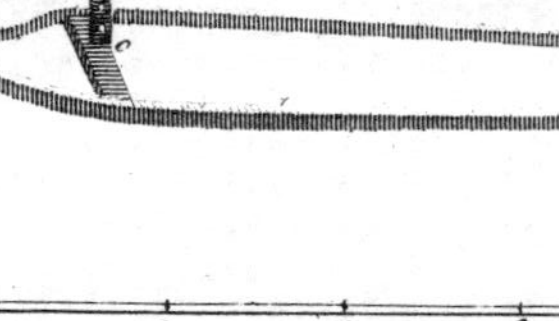
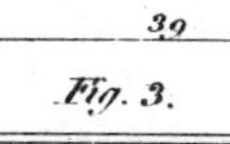
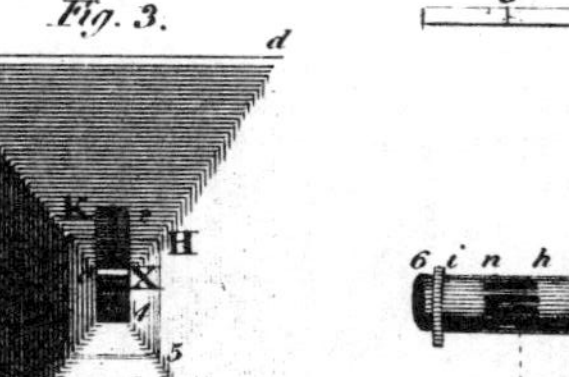

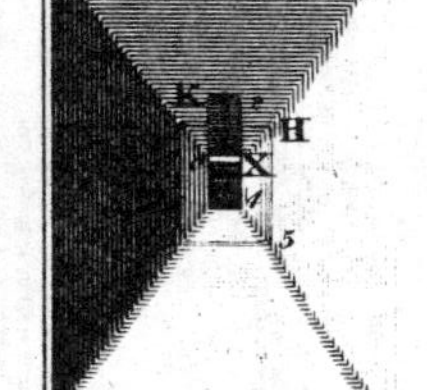

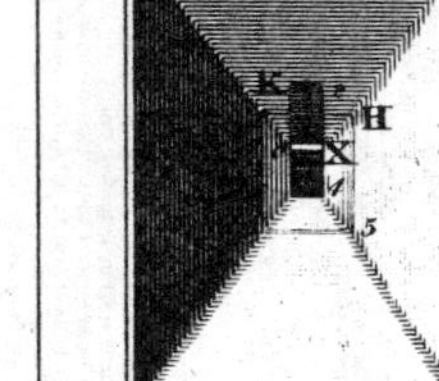

Oxford
A Single Drill.
Plate XXVII
Fig. 1.
1 2 3 4 5 6 Feet
Fig. 3.
Fig. 2.
3 6 9 12 Inches
Fig. 4.
one Inch
Hoole sc. Strand.

Oxford

Mr. Fane's Racks.
SIDE VIEW OF TURNIP RACK.
No. 1.

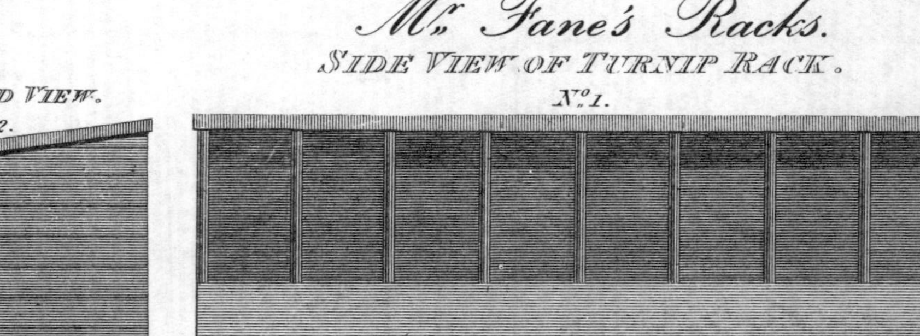

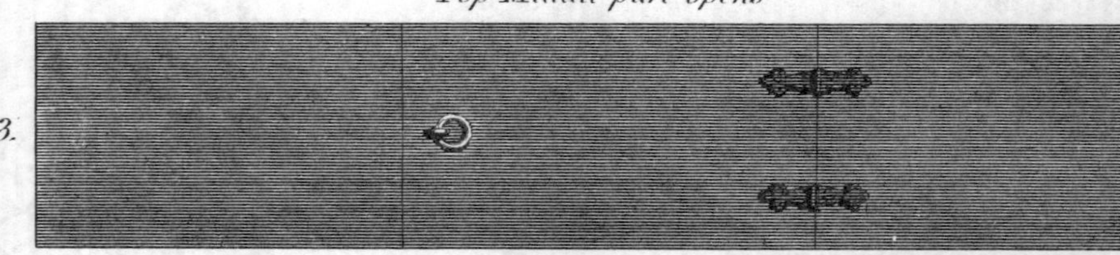

	Ft.	In.
Fig. 2.		
d 48,	4	0
3 24,	3	6
h 16,	1	4
h 24,	2	0

	Ft.	In.
Fig. 3.		
e 11,	0	11
e 5,	0	5
y z,	0	$6\frac{1}{4}$

	Ft.	In.
11 4,	0	$4\frac{1}{4}$
i k,	0	5
k z,	0	4
k n,	0	4

Plate XXVII. Single Drill.

Fig. 1, z N the screw to z K, the-regulator; c c, the cord to lift the wheel up when you do not want it to turn the seed out.

Fig. 2, the seed wheel and the seed casting cylinder; n h, the notches; e e, the seed.

Fig. 3, front view of the hopper, shewing part of the seed-casting notches; o x, the mouth where the seed passes; o K, part of the regulating tongue.

Fig. 4, a section of the seed-casting notches.

	Ft.	In.
Fig. 1.		
a b,	6	6
a 15,	1	3
b 39,	3	3
h 34,	2	10
h n,	2	0
h 4,	3	4
a x,	1	4
x 9,	1	2
9 18,	1	0
18 12,	1	6
x N,	0	10
N z,	0	7
z x,	0	11
a 5,	3	5
x 5,	2	1
5 p,	1	8
p 20,	1	8
b 6,	0	6
b i,	1	9

	Ft.	In.
i E,	2	7
o t,	1	0
i o,	1	3
e o,	1	6
o 8,	6	6
e 8,	8	0
Diameter.		
W,	2	0
Fig. 2.		
x 10,	0	10
x 19,	1	7
W 24,	2	0
19 6,	2	1
6 x,	0	6
6 i,	0	$1\frac{1}{2}$
i x,	0	$4\frac{1}{2}$
x h,	0	$1\frac{1}{2}$
n h,	0	$1\frac{1}{2}$
ee W,	1	6

	Ft.	In.
o x,	0	2
Fig. 3.		
12 d,	1	0
d 9,	1	6
9 18,	1	0
12 18,	1	6
3 5,	0	3
5 H,	0	5
4 2,	0	4
2 x,	0	2
x 4,	0	2
K 2,	0	1
o x,	0	1
Fig. 4.		
x i,	0	2
i o,	0	$0\frac{6}{16}$
i 4,	0	$0\frac{1}{2}$
t o,	0	$0\frac{6}{16}$

Plate XXVIII. Mr. Fane's Racks

No. 1. Is eight feet six inches long, one side two feet high,

high, the other two feet three inches, exclusive of the feet, which are two or three inches.

No. 2. The end view.

No. 3. The top, the middle part of which folds back on the rack. There is a bend on the lower side of the top, which serves as a gutter to prevent the rain-water falling from the rack on the sheep's back.

No. 4. The bottom, which opens the other way, and leans against the top, for a man to walk in to carry it.

THE END.

Printed by B. M‘Millan,
Bow Street, Covent Garden.